CARBONATE CHEMISTRY OF AQUATIC SYSTEMS: THEORY & APPLICATION

by

R.E. Loewenthal
Lecturer, Department of Civil Engineering
University of Cape Town

G.v.R. Marais
Professor of Water Resources and Public Health Engineering
University of Cape Town

ANN ARBOR SCIENCE
PUBLISHERS INC
P.O. BOX 1425 • ANN ARBOR, MICH. 48106

AUTHORS' NOTE

The incentive for the preparation of this monograph came from a desire to present a practical and useful exposition of the equilibrium chemistry of the Ca-Mg-Carbonic System. This system plays such a dominating rôle in softening and stabilization of water supplies, that a thorough understanding of its behavioural characteristics cannot but lead to improvement in the design, operation and control of water treatment processes.

The text is oriented to engineers and chemists, both student and practising. It deals thoroughly with the theoretical aspects, but wherever possible the presentation is amplified by graphical methods. Always the final objective was kept in mind - that the information must be in a practical, usable form, simple and quick to apply. The text includes numerous examples covering practically every problem likely to arise in evaluating waters and their modification by chemical dosing to achieve the desired end result. The major element in every solution is a graphical one. Practical engineers can acquire considerable facility in solving problems having only a qualitative knowledge of the basic chemistry.

This volume deals with the system characteristics in waters of low ionic strength (having less than about 1 000 mg/ℓ inorganic dissolved solids) where ion pairing can be neglected. The next volume will consider waters of high ionic strength, where ion pairing has a significant effect on the system characteristics. The procedures in this monograph apply only approximately to waters reclaimed from sewage because they contain phosphorus and organic materials. No consideration is given to boiler feed waters.

The investigation was supported by grants from the South African Council for Scientific and Industrial Research and the Water Research Commission of South Africa.

A special word of thanks is due to Mrs. D.F.Murcott who typed the manuscript for publication. Without her kindly good humour in dealing with two very temperamental writers it is doubtful if this monograph would have been completed.

The writers gratefully acknowledge the work of Mr. A.L.McLean who carried out the experimental verification reported in Chapter 6. They also wish to express their appreciation to Professor John Martin, Head of the Department of Civil Engineering, for his active support and encouragement.

R.E.Loewenthal,
Lecturer, Department of Civil Engineering,
University of Cape Town.

G.v.R.Marais,
Professor of Water Resources and Public
Health Engineering,
University of Cape Town.

iii

CONTENTS

Contents

Contents

xi

Chapter 1

INTRODUCTION

The earth's atmosphere contains approximately 0,03 per cent of
carbon dioxide by mass. Although such a small fraction, its
influence is as important to life on earth as is oxygen. Carbon
dioxide and oxygen pass through global cycles under the influence of energy input from the sun to participate intimately in
the production and destruction of life.

Life has developed on earth in the water phase and in general
can flourish only under very limited tolerances of certain phy-
sical and chemical properties within this phase. One important
property in this respect is the maintenance of the pH at around
7,0. Here carbon dioxide and its chemical compounds occupy a
central position of importance.

Apart from this aspect, carbon dioxide influences the
chemistry of waters profoundly. There is no doubt that an understanding of the interaction of the carbonic system on the
properties of water provides a powerful tool in the optimal
utilization of water by man.

The earth's atmosphere contains approximately 2000×10^9
tonnes of CO_2. The seas contain approximately $130\ 000 \times 10^9$
tonnes and rocks of the earth together with fossil fuels, approximately $40\ 000 \times 10^9$ tonnes [Plass, 1959].

Concentration of CO_2 in the atmosphere is maintained at an
approximate constant value, that of the equilibrium value established between dissolved CO_2 in the oceans and CO_2 in the
air. An annual exchange each way of approximately 100×10^9
tonnes of CO_2 takes place between the oceans and the atmosphere.

Fluctuations of CO_2 in the atmosphere are principally due
to photosynthesis in daylight hours, this fixes approximately
60×10^9 tonnes of CO_2 per year, and biological respiration
and decay which releases about the same amount. Without the

oceans it is very likely that fluctuations of CO_2 in the atmosphere would be pronounced - however the oceans serve as a giant source and sink to attenuate the fluctuations.

Since the industrial revolution massive quantities of CO_2 are being injected into the atmosphere from the burning of fossil fuels, approximately 6×10^9 tonnes per year. But because of the massive buffering capacity of the oceans it is not yet possible to state categorically whether there has been or shall be, in time, a significant change in the atmosphere.

When CO_2 dissolves in water it forms a weak acid. This acid and its salts in solution, called the carbonic system, form a buffer system which resists pH change. The pH of natural waters on land is controlled mainly by this system. The pH of the upper regions in the sea is controlled by the carbonic system and another weak acid system - that due to borate. In the lower regions of the sea pH appears to be controlled by the precipitation and solution of various minerals [Sillen, 1961].

In this monograph, consideration will only be given to the carbonic system in water where the concentration of total dissolved salts (T.D.S.) is less than about 1 000 ppm. Consequently the theory will find application principally in waters flowing over or within the land masses. For waters containing higher concentrations of dissolved salts the phenomenon of ion-pairing between cations (principally Ca^{++}, Mg^{++} and Na^{+}) and anions ($CO_3^=$, HCO_3^- and OH^-) becomes significant and the basic theory, though still valid, has to be extended to include this effect. A monograph is in preparation in which high salinity waters are considered.

In waters on land, the carbonic species are derived from a number of sources, i.e. CO_2 from the atmosphere dissolving into the water; CO_2 from biological oxidation of organic matter is liberated in water to form carbonic acid; water flowing over

carbonate-bearing sediments dissolves carbonate bearing minerals
into it.

The carbonic system in solution exists in four forms: dissolved CO_2, carbonic acid (H_2CO_3), bicarbonate (HCO_3^-) and carbonate
$(CO_3^=)$ ions. Their interaction with water results in a displace-
ment of equilibrium between H^+ and OH^- ions. This manifests
itself by the establishment of a specific pH. At this pH the
relative concentrations of H_2CO_3, HCO_3^- and $CO_3^=$ is fixed according to equilibrium reactions. It is usually convenient to relate these relative concentrations to pH, in this respect pH is
accorded the significance of an independent variable, whereas
it is the consequence of complex interaction between the dissolved species and water.

The relationship between dissolved CO_2 and H_2CO_3 is a parallel
reaction: in general there is always a fixed proportion (usually
about 0,3 per cent) of dissolved CO_2 forming H_2CO_3. If CO_2 is
stripped from or added to water the H_2CO_3 concentration changes
accordingly which results in changes in H^+, OH^-, HCO_3^- and $CO_3^=$
concentrations.

In addition to the dissolved carbonic species, all natural
waters contain dissolved and suspended substances in amounts
which depend on their previous history. When rain falls small
quantities of oxygen, ammonia, oxides of nitrogen and sulphur
are dissolved into the water. Impurities such as dust and bac-
teria are also picked up. As the rain water flows over ground
or percolates through soil, organic acids and mineral salts
such as iron, sodium, silicate are dissolved. Discolouration of
the water occurs due to dissolved and/or colloidally dispersed
organic matter. River waters receive these land washings and
also street drainage, treated effluents, raw sewage and all
manner of waste waters resulting from the activities of man.
These waters often have to serve as a source for (1) agricultural,

(2) industrial and (3) domestic water use.

Agriculture is the largest single user of water, principally irrigation. Minor farming uses are for watering purposes, care of livestock and poultry, and for cleaning.

The quality of irrigation water appears to be governed by four characteristics [Camp, 1963]:

(i)   Total concentration of soluble salts: excessive salinity at root-zone level leads in succession to leaf burn, leaf drop, twig dieback and plant destruction.

(ii) Relative proportion of sodium to other cations: if the sodium content of an irrigation water is high compared with the calcium and magnesium content, sodium will be adsorbed by the soil and replace the calcium and magnesium ions. High sodium concentration breaks down soil structures, seals pores and interferes with the drainage. Undesirable alkali conditions can develop in the soil.

(iii) Trace concentrations of certain elements (e.g. boron) are essential for plant growth; high concentrations of the elements, however, can be injurious to plants.

(iv) Concentration of bicarbonate ion: where there is high concentration of bicarbonate in a water there is a tendency for calcium and magnesium to precipitate as the solution becomes more concentrated in the root-zone. The relative amount of sodium in the water is thus increased - possibly to a level where conditions in paragraph (ii) are attained.

It is particularly apparent in (iv) that the carbonic system influences the suitability of agricultural waters. At present there is little that man can do to adjust by artificial means

the qualities of agricultural waters due to the large volumes
that are used daily. The only adjustment practical is that involving the carbonic system, e.g. the addition of $CaCO_3$ or
$Ca(OH)_2$ to soils. Here an understanding of the behaviour of the
carbonic system will result in improved evaluation of the effects
of such additions.

The quality of water required for industrial purposes depends
largely on the nature of the industry [James, 1940]. Thus many
industries have set up water standards for specific uses within
the industry. The use of 'hard' waters in the textile industry can
cause inferior products and uneven dyeing; water for laundering
industry is required to be free of discolouration forming contaminants and 'soft' enough to avoid soap wastage; in paper and
pulp industry 'hard' alkaline waters can cause problematic in-
terferences in precipitation processes; bacterial pollution of
water at certain stages in leather tanning can cause damage to
hides by putrefaction; boiler feed waters are required to be
conditioned to the state that they are neither scale forming
nor corrosive at high temperatures and pressures.

Water for domestic purposes is required to be free of taste,
colour and odour; free of biological contaminants; non-corrosive;
well buffered against pH change and of such a nature that it
easily forms a lather with soap. Hard waters are generally considered to be those that require considerable amounts of soap
to produce a lather, and also those that form a scale in units
where water temperatures are raised: e.g. boilers, kettles and
hot water pipes. Water for industrial and domestic purposes
can be treated to attain those qualities considered to be necessary for their efficient utilization. Such treatment intimately involves the carbonic system.

Hardness is caused by divalent metallic cations in solution:
e.g. calcium, magnesium, strontium, ferrous and manganous ions.

These divalent cations are associated with a number of anions in solution: $CO_3^=$, HCO_3^-, $SO_4^=$, Cl^-, $SiO_3^=$. Calcium and magnesium are the major contributors to hardness in natural waters. It is usually assumed in practice that the concentrations in solution of these two cations constitute the *total hardness* of the water. That part of total hardness which is chemically associated with $CO_3^=$ and HCO_3^- ions in the water is called *carbonate hardness* and the concentration of hardness which is in excess of the carbonate hardness is called *noncarbonate hardness*.

Depending on the pH concentrations of calcium ions and carbonic species a water may have a tendency either to precipitate calcium carbonate ($CaCO_3$) out of solution, or to dissolve solid $CaCO_3$ into solution. Such waters are termed *'oversaturated'* and *'undersaturated'* (or aggresive) respectively.

Water treatment to attain the desired qualities is carried out in a series of unit processes: chemical flocculation, filtration, softening, stabilization and chlorination. The efficiency of each unit process is, in part, dependent on carbonic species concentration, pH and hardness: the adjustment of these parameters to desired values, using chemical additives, is called *conditioning*. Conditioning a water for the removal of hardness is called *softening*. Conditioning a water so that it is 'non-corrosive' is called *stabilization*.

Present day practice for estimating the mass of chemical additives in conditioning is based principally on stoichiometric relationships. This method of estimation ignores fundamental ionic and solubility equilibrium chemistry, as well as temperature and ionic strength effects on chemical reactions. Because stoichiometric calculations are relatively simple this method has attained widespread use. However, dosage determinations by this method often need to be adjusted on the basis of past experience. For example, the term 'excess lime treatment' is in

part due to the inability of the stoichiometric method to predict with reasonable accuracy the required lime dosage for
hardness removal.

The objective of this monograph is to present an equilibrium
approach to the chemistry of the carbonic system in natural and
treated waters. It is a relatively complicated system and one
cannot reasonably expect its straight application to every day
problems in water treatment. Fortunately it is possible to
apply the method readily and in a simple manner by developing
graphical aids. These graphical aids are most illuminating in
bringing an insight into the behaviour of the carbonic system.
Once one attains this understanding it is a relatively simple
matter to transfer this knowledge to other chemical problems in
water treatment. For example, aspects of chlorination chemistry, particularly the influence of pH, becomes clearer not only
qualitatively but also quantitatively. In a similar fashion
the elementary aspects of the chemistry of phosphates, silica
and ammonia in water are more readily appreciated.

Chapter 2 will deal with the basic theory of ionic equilibrium chemistry. In Chapter 3 the theory is applied specifically to the carbonic system. In Chapter 4 solubility equilibrium between the solid phase of a substance and its dissolved
phase is considered. In Chapter 5 graphical methods are developed for estimating chemical dosages in:

1. Single Phase Model, which considers only equilibrium
    between dissolved carbonic species.

2. Two Phase Model, which considers equilibrium between
    the dissolved phase of the carbonic species and either
    (i) solid phase ($Ca(CO_3)$ and $Mg(OH)_2$), or (ii) gas
    phase (CO_2 in the air). Application to softening prob-
    lems is considered in detail.

3. Three Phase Model, which considers equilibrium between
    dissolved species, solid phase ($CaCO_3$) and gas phase
    (CO_2 in the air).

   Chapters 6 and 7 deal respectively with the conditioning of
soft waters for municipal use and conditioning of water for high
temperature (hot water) systems.

References

Camp, T.R.   1963   *Water and its Impurities*, Reinhold
                            Publ. Corp., New York.

James,G.V.   1940   *Water Treatment*, Technical Press,
                            London.

Plass,G.N.   1959   "Carbon Dioxide and Climate", Sic.
                            Amer., (July), 3-9.

Chapter 2

BASIC CHEMISTRY OF LOW IONIC STRENGTH WATER

Chemical Equilibrium

Chemical equilibrium is a dynamic state of balance between reactants and reaction products. The opposing rates of reaction,
forward and reverse, are exactly equal.

A reaction is represented as

$$A + B \rightleftharpoons C + D.$$

The law of mass action states that for an 'ideal solution' the
rate of generation of products C and D is proportional to the
product of the molar concentrations of reactants A and B, i.e.

$$v_f = k_f \, [A][B]$$

v_f = velocity of forward reaction

k_f = constant of proportionality

[ ]   indicates molar concentration[1]

The rate of the reverse reaction (i.e. the reforming of reactants A and B) is subject to the same law: the rate is proportional to the product of the molar concentrations of the
reaction products C and D, i.e.

$$v_r = k_r [C][D]$$

where

v_r = velocity of reverse reaction

k_r = constant of proportionality.

---

1. The molar concentration of a substance is the number of
gram moles of that substance in a litre of solution. One gram
mole per litre of a substance is its molecular weight in grams
in one litre of solution

At equilibrium $v_f = v_r$, i.e.

$$k_f[A][B] = k_r[C][D]$$

therefore

$$\frac{[C][D]}{[A][B]} = \frac{k_f}{k_r} = K \qquad\qquad\qquad (1)$$

Equation (1) is an example of an equilibrium equation. K is the equilibrium constant. Its value for a particular reaction changes with temperature and the ionic strength of the solution.

In the theoretical development set out here the assumption is made of 'ideal solutions' (no ionic strength effects) at 25°C. The influence of temperature and ionic strength on K is discussed later in this chapter.

It is important to note that in Eq. (1) the molar concentrations [A], [B], [C] and [D] are the concentrations in solution at equilibrium. Usually the initial concentration will differ from the equilibrium concentration.

Ionic Equilibrium

In solution many substances split apart in charged species, dissociating to form positively and negatively charged ions. The dissociation process is called *ionization*, the positively charged ion - the cation and the negatively charged ion - the anion.

The equilibrium reaction for substance AB which ionizes only partly in solution can be written as:

$$AB \mathrel{\substack{\leftarrow \\ \rightarrow}} A^+ + B^-$$

Applying the law of mass action to the above equilibrium reaction:

$$[A^+][B^-]/[AB] = K$$

where K is the ionization (dissociation) constant of substance AB.

Acids and Bases

According to the Arrhenius interpretation, an acid is a substance which yields hydrogen ions, H^+, on dissociation; a base is a substance which yields hydroxyl ions, OH^-, on dissociation. For many purposes the Arrhenius interpretation is adequate, and it is a concept easily appreciated by non-chemists. In this monograph the Arrhenius interpretation will be used in preference to the more general one by Lewis.

Ionization of Water

Interpreted in terms of an acid or base, water possesses the property of being both at the same time. The consequences of this unusual property makes it necessary to enquire more closely into the ionization of water.

The ionization of water is conventionally written as:

$$H_2O \rightleftharpoons H^+ + OH^-$$

Equilibrium constant for this reaction is

$$\frac{[H^+][OH^-]}{[H_2O]} = K \qquad (2)$$

where K is the ionization constant for water at 25°C.

From experimental work K is found to be an extremely small quantity, $1,8 \times 10^{-16}$ moles/litres at 25°C. From Eq. (2):

$$[H^+][OH^-] = K[H_2O] \qquad (3)$$

and as K is small the fraction of H_2O that ionizes is negligible compared with the unionized fraction. The unionized mass of H_2O can be taken as equal to the total water mass,

$$[H_2O] = \frac{\text{mass of 1 litre of water}}{\text{gram molecular weight of water}}$$

$$= \frac{1000}{18} = 55,5 \text{ g moles/litre}$$

$$[H^+][OH^-] = K \times [H_2O]$$

$$= 1,8 \times 10^{-16} \times 55,5$$

$$= 1,0 \times 10^{-14}$$

i.e.

$$[H^+][OH^-] = K_w = 10^{-14} \text{ (at 25°C)} \qquad (4)$$

The value of K_w is a function of temperature and ionic strength, that is, the total concentration of ions in the water. The value given above is correct only for very dilute aqueous solutions at 25°C.

In pure water:

$$[H^+] = [OH^-]$$

from Eq. (2)

$$[H^+] = [OH^-] = \sqrt{K_w}$$

$$= 1,0 \times 10^{-7} \text{ moles/litre at 25°C.}$$

pX Notation

Where small numbers are involved it is convenient to express concentrations or equilibrium constants in terms of logarithms instead of the arithmetic values. These are given the general notation 'pX' where for an 'ideal solution':

$$pX = -\log_{10}[X]$$

and [X] is a concentration in moles/litre. For example:

$$pH = -\log_{10}[H^+]$$
$$= \log_{10}(1/[H^+])$$

i.e.

$$[H^+] = 10^{-pH} \ .$$
(5)

For pure water at 25°C:

$$[H^+][OH^-] = K_w = 10^{-14}$$

$$\log_{10}([H^+][OH^-]) = \log_{10}K_w$$

$$\log_{10}[H^+] + \log_{10}[OH^-] = \log_{10}K_w$$

$$-pH - pOH = -pK_w$$

$$pH + pOH = pK_w = \log_{10}10^{14} \ .$$

For a neutral solution at 25°C:

$$[H^+] = [OH^-]$$

$$[H^+]^2 = 10^{-14}$$

$$[H^+] = 10^{-7}$$

$$pH = pOH = 7,0 \ .$$

Weak and Strong Acids and Bases

Acids and bases that dissociate completely, or almost completely, in solution are called 'strong'. Those that ionize only partially in solution are called 'weak'. Irrespective of the degree of dissociation, the following applies:

Equilibrium reactions and equations for the dissociation of an acid, HA, and a base, BOH, are:

Acid: $HA \rightleftharpoons H^+ + A^-$
(6)

$$\frac{[H^+][A^-]}{[HA]} = K_a$$
(7)

(8)

Base: $BOH \rightleftharpoons B^+ + OH^-$

$$\frac{[B^+][[OH^-]}{[BOH]} = K_b \tag{9}$$

K_a and K_b are the dissociation constants for acids and bases respectively and are usually listed as such in the literature.

In aqueous solutions a base can always be interpreted as an acid (and vice versa). It is a matter of convenience which form one chooses. For our purposes the more convenient is the acid form for the reason that the equilibrium equation incorporates H^+ which relates directly to pH - a parameter normally measured.

The equilibrium reaction for base BOH can be written in its acid form as follows:

$$B^+ + H_2O \leftrightharpoons BOH + H^+$$

i.e.

$$\frac{[BOH][H^+]}{[B^+]} = K_{ab} \tag{10}$$

where K_{ab} is the acid equilibrium constant for the base BOH.

The equilibrium constant for the acid form of a base, K_{ab}, is related to the base form, K_b, as follows: solving for $[OH^-]$ from Eq. (4) and substituting this value into Eq. (9) and simplifying:

$$\frac{[BOH][H^+]}{[B^+]} = \frac{K_w}{K_b} = K_{ab}.$$

For convenience K_{ab} is written as K_a. In this monograph, except where stated otherwise, all equilibrium equations will be written using the acid form of the equilibrium constant. Consequently the subscript 'a' is often omitted.

Interpreting all acids and bases as acids, three classes of acids can be identified. Those with:

(i) K_a greater than one - giving pK_a less than zero.
These are strong acids which are considered as
completely dissociated in the pH range 0 to 14;

(ii) pK_a between 0 and 14 - these are weak acids and
are only partly dissociated in the range pH
0 to 14;

(iii) pK_a greater than 14 - these are very weak acids
and correspondingly strong bases which are considered completely dissociated in the range pH
0 to 14.

Multi-protic Acids and Bases

Depending on the particular acid or base, acids and bases may
yield one or more hydrogen or hydroxide ions on dissociation.
An acid capable of liberating one hydrogen ion on dissociation
is called monoprotic and will have one ionization constant; two
hydrogen ions - diprotic with two ionization constants, etc.

Carbonic acid is an example of a weak diprotic acid and is
thus characterized by two ionization reactions,

$$H_2CO_3 \leftrightarrows H^+ + HCO_3^-$$

and

$$HCO_3^- \leftrightarrows H^+ + CO_3^=$$

These equilibrium reactions yield the following equilibrium
equations

$$\frac{[H^+][HCO_3^-]}{[H_2CO_3]} = K_1$$

and

$$\frac{[H^+][CO_3^=]}{[HCO_3^-]} = K_2$$

Values of K_1 and K_2 for weak acids and bases usually differ by approximately two or more orders of magnitude.

Addition of Strong Acids and Bases to Water

If C_T moles/litre of a strong monoprotic acid is added to water, complete ionization of the acid occurs, i.e.

$$HC\ell \rightarrow H^+ + C\ell^-$$

and, because the only source of $C\ell^-$ in the water is HCl

$$C_T = [C\ell^-] \tag{11}$$

The addition of H^+ to the water causes an increase in the equilibrium concentration of $[H^+]$ and a corresponding decrease in $[OH^-]$ so that the equilibrium equation for water, Eq. (4) be satisfied.

To estimate the pH established in the water, note that Eq. (4) incorporates two unknown parameters; thus a further independent relationship is needed for a unique solution. Such a relationship can be derived by considering either a mass balance on the hydrogen ions or an electron charge balance.

(i) Mass balance on $[H^+]$: The *change* in equilibrium concentration of H^+ before and after strong acid addition, equals the strong acid added minus that H^+ re-combining with OH^- so that Eq. (4) be satisfied:

$$\Delta[H^+] = C_T - \Delta[OH^-]$$

i.e.

$$[H^+]_2 - [H^+]_1 = C_T - ([OH^-]_1 - [OH^-]_2) \tag{12}$$

where subscripts 1 and 2 refer to equilibrium concentrations before and after addition of strong acid respectively. Furthermore, because the solution was initially pure water $[H^+]_1$ equals $[OH^-]_1$. Substituting $[H^+]_1 = [OH^-]_1$ in Eq. (12)

gives

$$[H^+]_2 = C_T - [OH^-]_2 \ . \tag{13}$$

(ii) Electron charge balance equation: The solution must be electro-neutral; the sum of the positive charges must equal the sum of the negative charges:

$$\sum_x^x (N_x[X^{+N_x}]) = \sum_y^y (M_y[Y^{-M_y}])$$

$[X^{+N_x}]$ = molar concentration of cation species X
            with charge N_x

$[Y^{-M_y}]$ = molar concentration of anion species Y
            with charge $-M_y$.

For the strong acid HCℓ in pure water the only charged species in solution are H^+, OH^- and $C\ell^-$. Balancing the concentrations of negative and positive charges

$$[H^+] = [C\ell^-] + [OH^-].$$

Substituting for $C\ell^-$ from Eq. (11),

$$[H^+] = C_T + [OH^-]$$

which is the same as Eq. (13).

The unique solution for $[H^+]$ can now be derived. Solve for $[OH^-]$ in Eq. (4), substitute for $[OH^-]_2$ in Eq. (13): solving for $[H^+]_2$ gives

$$[H^+]_2 = \frac{C_T \pm \sqrt{C_T^2 + 4K_w}}{2} \tag{14}$$

Equation (14) shows that if the molar mass of strong acid added is somewhat greater than $2\sqrt{K_w}$, i.e. 2×10^{-7} moles/ℓ (even as small as 2×10^{-6} moles/ℓ), the hydrogen ion concentration is closely given by the molar mass of strong acid added thus:

$$pH = pC_T \quad .$$ (15)

Addition of Weak Acids and Bases to Water

If a monovalent weak acid, HA, is added to water, partial ionization of the acid occurs:

$$HA \rightleftharpoons H^+ + A^-.$$

Assuming an ideal solution, the extent of ionization depends on
the ionization constant for the particular acid:

$$\frac{[H^+][A^-]}{[HA]} = K_a$$

where [HA] is the concentration of undissociated acid in the
water at equilibrium.

In Eq. (15) it was shown that for strong acids (which
ionize completely) the concentration of H^+ in solution equals
the molar mass of acid added. Weak acids and bases ionize only
partially. In consequence the increase in H^+ or OH^- concentrations is not equal to the molar mass of weak acid or base added.
The problem becomes one of interdependence between ionization
of the weak acid and ionization of water.

Suppose C_T moles/litre of weak acid, HA, are added to water,
two interrelated reactions will occur:

dissociation of HA

$$HA \rightleftharpoons H^+ + A^-$$

dissociation of water

$$H_2O \rightleftharpoons H^+ + OH^-$$

Idealized equilibrium equations for these two reactions are:

$$\frac{[H^+][A^-]}{[HA]} = K_a \quad .$$ (16)

Where [HA] is the concentration of unionized acid in solution.

$$[H^+][OH^-] = K_w \quad . \tag{17}$$

Equations (16 and 17) incorporate four parameters - $[HA]$, $[A^-]$, $[H^+]$ and $[OH^-]$. Hence four independent relationships are required for a unique solution. Equations (16 and 17) supply two relationships, therefore two further independent relationships must be established.

These can be established by considering (1) a mass balance on the total weak acid species in solution and (2) an electron charge balance.

(1) Species balance. Mass balance on the concentrations of weak acid species in solution gives

$$C_T = [HA] + [A^-] \tag{18}$$

(2) Electron charge balance. For the weak acid HA in pure water, the only charged species in solution are $[H^+]$, $[OH^-]$ and $[A^-]$. Balancing the concentrations of negative and positive concentrations:

$$[H^+] = [OH^-] + [A^-] \tag{19}$$

The equilibrium concentrations of $[H^+]$, $[OH^-]$, $[A^-]$ and $[HA]$ are then solved from the four independent relationships Eqs. (16 to 19).

Besides the two independent relationships developed above, other pairs of such relationships exist. The parameters in one pair can always be expressed in terms of the parameters of any other pair. The pair selected is usually either one that contains parameters convenient for theoretical purposes, or one that contains parameters which can be measured. The pair selected above are convenient for the theoretical development of the system. However, when dealing with the practical aspects of the carbonic system, the most convenient independent pair is usually pH and alkalinity, primarily because they can both be measured relatively easily.

Addition of Weak Diprotic Acids

If C_T moles/litre of a weak diprotic acid, H_2A, is added to water the interaction of three dissociation reactions (two for the acid and one for water) are involved at equilibrium,

$$H_2A \leftrightharpoons HA^- + H^+$$

$$HA^- \leftrightharpoons A^= + H^+$$

$$H_2O \leftrightharpoons OH^- + H^+$$

Equilibrium equations to be simultaneously satisfied for these reactions are

$$\frac{[HA^-][H^+]}{[H_2A]} = K_1$$

$$\frac{[A^=][H^+]}{[HA^-]} = K_2$$

$$[H^+][OH^-] = K_w$$

These three equations contain five unknowns, $[H_2A]$, $[HA^-]$, $[A^=]$, $[H^+]$ and $[OH^-]$, hence two more relationships are required. These are found from the total species concentration of the acid and electro-neutrality of the solution.

The total acid species in solution is a constant and its sum must equal the concentration of weak acid added,

$$C_T = [H_2A] + [HA^-] + [A^=].$$

For electro-neutrality of the solution:

$$[H^+] = [OH^-] + [HA^-] + 2[A^=].$$

The solution of the above five equations leads to third and higher order algebraic equations. With slightly more complicated systems the analytical solution rapidly becomes untenable for practical purposes. A different approach is required to bring about quick practical solutions. Such an approach was first suggested by Bjerrum (1914) and Sillen (1959). They showed that a generalized solution in terms of pH can be rapidly drawn graphically and the particular solution found by imposing conditions appropriate to the problem. This method of solution is both useful and illuminating and bears detailed investigation.

Graphical Presentation of Acids and Bases

Suppose C_T moles/litre of weak acid HA is added to pure water. Dissociation of the acid and water gives:

$$HA \leftrightharpoons H^+ + A^-$$

$$H_2O \leftrightharpoons H^+ + OH^-$$

Equilibrium equations for these two reactions are:

$$\frac{[H^+][A^-]}{[HA]} = K_a \tag{20}$$

$$[H^+][OH^-] = K_w \tag{21}$$

These equations give two independent relationships for a system with four unknowns. Mass balance for the weak acid species in solution gives a third relationship:

$$C_T = [HA] + [A^-] \tag{22}$$

Only one more independent relationship is required. If the pH is known, or measured, $[H^+]$ is known and sufficient relationships exist for a unique solution. If pH is not known, then each of the species concentrations can be calculated for a series of pH values and a diagram drawn of species concentrations versus pH. This diagram then represents a general solution of the system in terms of pH.

To obtain the equations for plotting this species-pH diagram, each of the species concentrations is solved in terms of C_T and/or $[H^+]$ using Eqs. (20 to 22):

$$[H^+] = 10^{-pH} \tag{23}$$

Solve for $[A^-]$ in Eq. (22) and substitute into Eq. (20). Solving for $[HA]$ gives

$$[HA] = \frac{C_T [H^+]}{(K_a + [H^+])} \tag{24}$$

Substituting for $[HA]$ from Eq. (24) into Eq. (20) and solving for $[A^-]$ gives

$$[A^-] = \frac{C_T K_a}{(K_a + [H^+])} \tag{25}$$

Using Eqs. (23 to 25), for a fixed value of C_T, each of the concentrations $[H^+]$, $[OH^-]$, $[HA]$ and $[A^-]$ are plotted against pH.

A more useful form is to plot the logarithm of each of the species concentrations against pH. Taking the logarithm of Eqs. (24 and 25):

$$\log [HA] = \log C_T - pH - \log (K_a + [H^+]) \tag{26}$$

$$\log [A^-] = \log C_T - pK_a - \log (K_a + [H^-]) \tag{27}$$

and by definition

$$\log [H^+] = -pH \tag{28}$$

i.e.

$$\log [OH^-] = pH - pK_w \tag{29}$$

Knowing C_T, for each pH calculate $[H^+]$ from Eq. (25) and substituting into Eqs. (26 to 29) the parameters log

[HA], [A⁻], [H⁺] and [OH⁻] are plotted versus pH. An example
of such a plot is shown in Figure 1 from which it will be seen
that:

(i) in the region pH << pK_a the [HA] line is essentially
horizontal with ordinate value equal to C_T; [A⁻] line
is approximately 45°, with slope + 1. Projecting the
straight sloping section of the [A⁻] line back it intersects the C_T line at pH = pK_a,

(ii) where pH = pK_a the [HA] and [A⁻] lines intersect
(i.e. [HA] = [A⁻]). Substituting in Eqs. (6 and 7)
and solving gives [HA] = [A⁻] = 1/2 C_T i.e. log [HA]
= log [A⁻] = log ([C_T]/2) = log C_T - 0,303, i.e. the
point of intersection is 0,303 units below the C_T
line,

(iii) where pH >> pK_a the [A⁻] line is essentially horizontal and [HA] is approximately at 45°, with slope - 1.
Projecting the straight sloping section of the [HA]
line it intersects the C_T line at pH = pK_a.

These observations lead to a rapid approximate method of
constructing log [species] - pH diagrams (Bjerrum, 1914; and
Sillen, 1959):

(i) Draw axes for pH and log [species] with equal scales
(see Figure 1) starting from pH = 0 to pH 14 and
log [species] = 0 to -14.

(ii) Plot the C_T value on the log [species] axis and draw
a horizontal line through this point.

(iii) Plot the pK value on the pH axis and draw a vertical
line through pK to intersect the C_T line. The intersection defines the 'helping point', 1, in Figure 1.

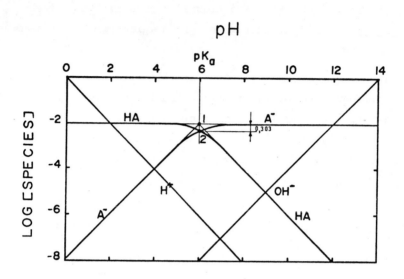

Figure 1. Distribution of species with pH for a monoprotic acid; $C_T = 10^{-2}$M and $pK_a = 6,0$

(iv) Through the helping point draw lines at 45° below the C_T line with slopes +1 and -1.

(v) Plot Point 2 in Figure 1, 0,303 log units vertically below helping Point 1.

(vi) Sketch two curves through Point 2 within about 1 pH unit on each side of the pK value such that each curve links a 45° line with the horizontal C_T line on the opposite of the pK value.

(vii) Draw [H^+] line at 45° with slope -1 to pass through the pH axis at zero, log [species] at zero.

(viii) Draw [OH^-] line at 45° with slope +1 through the pH axis at 14 and log [species] axis at zero. The [H^+] and [OH^-] lines intersect at the coordinate point pH = 7 and log [species] = -7.

The log [species] - pH diagram for a polyprotic acid can be
plotted using similar techniques to those set out above for
monoprotic acids.   The method uses the assumption that the
polyprotic acid can be represented by a number of monoprotic
acids. *Each* of these monoprotic acids has a C_T value equal to
that for the polyprotic acid and a pK value equal to the successive pK values for the polyprotic acid.   Lines representing
the species for each pK value are plotted as set out above for
a monoprotic acid. The horizontal lines representing species
common to successive dissociations are then joined. Example:
sketch the log [species] - pH diagram for a 0,01 M diprotic
acid, H_2A, with log dissociation constants pK_1 and pK_2 equal to
6,33 and 10,30 respectively. The distribution of species for
K_1, H_2A and HA^-, are first plotted about pK_1 with C_T equal to
0,01 (see Figure 2). The species lines for HA^- and $A^=$ are then
plotted about pK_2 again with C_T equal to 0,01 M.   The common

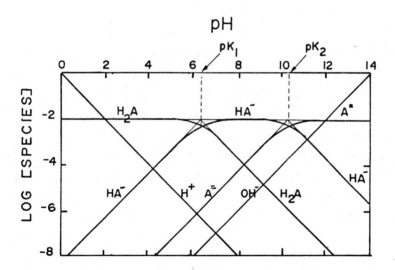

Figure 2. Log [species] - pH diagram for a diprotic acid,
$C_T = 10^{-2}$M

[HA$^-$] line is then joined.

The shape of the log [species] – pH diagram is independent of the total species concentration, C_T. As C_T is increased or decreased, the whole diagram moves vertically up or down accordingly, but the shape remains the same. This is illustrated in Figure 3.

At any particular pH in this diagram log $[C_T]_1$ – log $[C_T]_2$ = log $[HA]_1$ – log $[HA]_2$ = log $[A^-]_1$ – log $[A^-]_2$. Thus:

$$\frac{[C_T]_1}{[C_T]_2} = \frac{[HA]_1}{[HA]_2} = \frac{[A^-]_1}{[A^-]_2} \ .$$

We have stated that the pH – log [species] diagram represents a general solution in terms of pH. In any particular problem, additional information must be supplied to obtain a unique solution; either the pH must be given or the mass addition of acids and bases (both strong and weak) and their salts, must be specified.

Apart from providing a means of obtaining particular solutions to chemical equilibrium problems, the diagram serves as a unifying agent in the development of theoretical concepts of strong and weak acids and bases, mixtures of weak and strong acids and bases, alkalinity and acidity, buffer capacity and pH – titration curves. These entities are well established in aqueous chemistry but their close interrelationships are often not so apparent from their formulation.

Before entering into a detailed development of the entities above, one problem needs clarification. It was shown that for a unique solution to acid-base systems the following equations are needed:

(1) Equilibrium equations

(2) Two additional independent relationships.

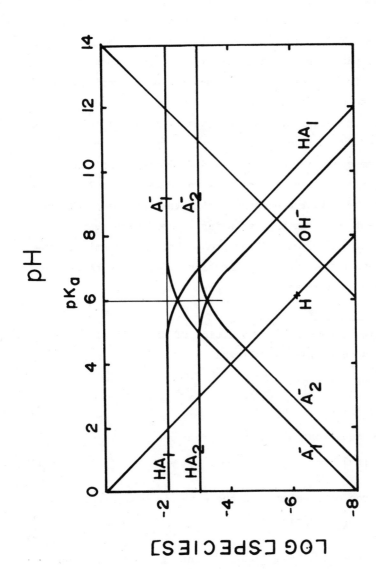

Figure 3. Effect of C_T on distribution of species with pH;
$C_T = 10^{-2}$M, $C_T^2 = 10^{-3}$M

It was further shown that one of the independent equations -
mass balance on acid species in solution - could be integrated
with the equilibrium equations to give a general solution in
terms of a second independent parameter, pH, in the log [species]
- pH diagram. To obtain a unique solution if pH is unknown, a
second independent relationship is needed. Up to now a charge
balance equation was used. This approach was convenient while
the ions in solution were associated with only the acid and
base species in solution. When dealing with normal terrestrial
waters there may be many additional ionic types in the water
which in themselves do not directly contribute to shifts in
equilibrium except in so far as they increase the ionic
strength of the solution - consequently they can be eliminated.
An independent relationship replacing the electron charge
balance relationship, called the proton balance equation, containing only species directly affecting pH will now be derived
from the mass balance and charge balance equations.

Proton Balance Equation

A proton balance equation for addition of the salt of a weak
acid, NaA, is developed as follows:

Charge balance equation:

$$[Na^+] + [H^+] = [OH^-] + [A^-] \tag{30}$$

Mass balance equation:

$$C_T = [HA] + [A^-] \tag{31}$$

The only source of species A in the water is NaA, hence:

$$[Na^+] = C_T = [HA] + [A^-] \tag{32}$$

Substituting for $[Na^+]$ from Eq. (32) into Eq. (30) and simplifying:

$$[HA] + [H^+] = [OH^-] \tag{33}$$

which is the required proton balance equation.

Establishment of the proton balance equation by the procedure above is general but tedious - rapid establishment of this equation is a great convenience in theoretical and practical work. For this reason the following empirical procedure developed from the work of Bard (1966), involving the concept of proton levels, is presented. The zero proton species level is defined by the species *before* addition to water. For example, if NaA is added to water the zero proton species level is at NaA and H_2O. The species formed in solution are Na^+, A^-, HA, H^+ and OH^-. Relative to NaA (the zero level) the weak acid species A^- and HA gain 0 and 1 hydrogen ions repectively. These changes can be depicted as shown in Table 1. The water is assumed to gain and lose hydrogen ions relative to the zero condition (H_2O) giving H_3O^+ (i.e. H^+)and OH^-, see Table 1.

Table 1.

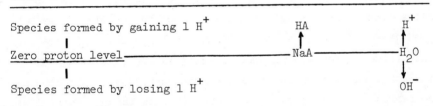

The proton balance equation is now written by equating the sum of the hydrogen ions gained to the sum of those lost by the species:

$$[HA] + [H^+] = [OH^-]$$

which is the same as Eq. (33).

pH Established by Addition of Strong Acids or Bases

The pH established by addition of strong acid or base to pure
water is best illustrated by means of an example.

Example 1. 10^{-2} moles/litre of strong acid, HCl, is added
to pure water; determine the pH. Proton balance equation is
developed from the proton levels set out below:

Species formed by gaining 1 H$^+$

Zero proton level ———————————— HCl ———————— H$_2$O

Species formed by losing 1 H$^+$ Cl$^-$ OH$^-$

giving:

$$[H^+] = [Cl^-] + [OH^-] \tag{34}$$

The log [species] - pH diagram for this system is shown in
Figure 4a. A line representing the sum of species on the right
hand side of Eq. (34), line A, is sketched in Figure 4a as follows: At high pH values [Cl$^-$] is negligible compared with [OH$^-$]
and the line representing their sum follows the OH$^-$ line. At
the pH were [Cl$^-$] equals [OH$^-$], Point 1 with pH 12,0, the ordinate value of the sum of these species is log 2 (i.e. 0,303)
units above the intersection point. The curved section of the
line representing this sum is sketched to pass through Point 1
and join up with the straight line section at one pH unit on
each side of Point 1. The line representing the left hand side
of Eq. (34), Line B, follows the [H$^+$] line. Graphical solution
to Eq. (34) is given by the intersection point of Lines A and B
that is, pH 2,0.

Example 2. 10^{-2} moles/litre of a strong base (i.e. a very weak
acid with pK$_a$ greater than 14) is added to water; determine the
pH.

Proton balance equation for the system is

$$[B^+] + [H^+] = [OH^-] \tag{35}$$

The left and right hand sides of this equation are represented
by Lines A and B respectively in Figure 4b; their intersection
point gives the pH established in the water as pH 12,0

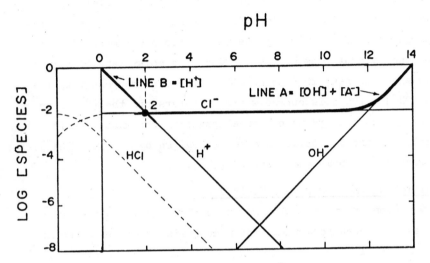

Figure 4a. Addition to pure water of 10^{-2}M strong acid; pH = 2,0

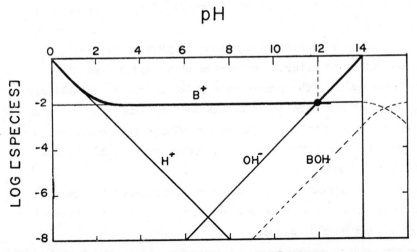

Figure 4b. Addition to pure water of 10^{-2}M strong base; pH = 12,0

The plots on the log [species] - pH diagram give a graphical
verification of the equations for pH for addition of a strong
acid or base. When the molar mass of acid added (for example)
is greater than about 10^{-6} moles/litre, the pH is given by the
intersection of the two straight lines for $[C\ell^-]$ and $[H^+]$. This
implies that pH is given by the log of the molar mass of acid
added as evident from Eq. (14). When the molar mass of strong
acid added is near 10^{-7} moles/litre, the line representing the
sum $[OH^-] + [C\ell^-]$ is curved where it intersects the $[H^+]$ line.
This implies that pH is no longer given by the mass of acid
added, and the dissociation of water (given by the term $[OH^-]$)
becomes of the same magnitude as $[C\ell^-]$.

Addition of Weak Acids and Their Salts

It was shown that dissociation is complete when a strong acid
or base is added to water. The hydrolysis effect of water is
normally negligible. In consequence the pH can be very simply
calculated; for example in the addition of a strong acid the
pH is equal to the negative log of the molar mass of acid
added. A weak acid is only partially dissociated in the range
pH 0 to 14. Hence the pH is *not* directly given by the molar
mass of acid added.

To introduce a consistent approach to pH when weak acids or
bases are added to water, a reference pH is defined as the
equivalence point. The equivalence point is the pH established
by addition of say x moles of a weak acid or base to a litre of
pure water. The x moles per litre constitute the *equivalent
solution*. Note that the equivalence point depends on the mass
of weak acid added.

The salt of a weak acid in solution dissociates to yield A^-
ions, hence there is also an equivalence point for addition of
salt of a weak acid to water.

The concept of equivalence points and equivalent solutions

is very useful when working with mixtures of weak and strong
acids and bases; it enables one to separate out a mass of strong
base (or acid) associated with a mass of either a weak acid (or
base) or a weak salt. These masses of weak acids (or base) or
weak salts are expressed as equivalent solutions.

Equivalence points for weak acids and their salts can be
graphically determined by solving the relevant proton balance
equation in the log [species] - pH diagram. The method is illustrated for mono- and diprotic acids and their salts.

1. Monoprotic Acid Systems

(i) Addition of a weak monoprotic acid: 0,01 moles/litre of
acid HA is added to water, estimate the pH. $pK_a = 6,0$. The
proton balance equation is established from the zero proton
condition as follows:

Species formed by gaining 1 H^+

Zero proton condition ——————————————— HA ——————————— H_2O

Species formed by losing 1 H^+ A^- OH^-

The required equation is thus

$$[A^-] + [OH^-] = [H^+] \tag{36}$$

The log [species] - pH diagram for this system is shown in
Figure 5. Referring to this diagram, the left and right hand
sides of Eq. (36) are effectively represented by the Lines A
and B respectively and their intersection point is at Point 1.
The pH value of this point is the pH established in the solution, i.e. pH 4,0, which is the equivalence point for a 0,01 M
solution of this weak acid.

(ii) Addition of the salt of a weak monoprotic acid: 0,01 moles
/litre of the salt NaA is added to water. Determine the equivalence point. Assume $pK_a = 6,0$.

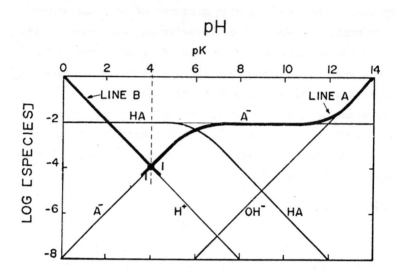

Figure 5. Addition of 10^{-2} moles/ℓ HA to pure water (equivalent HA solution)

Proton balance equation for this addition is given by Eq. (33). The log [species] - pH diagram for the system is shown in Figure 6. Referring to this diagram: the left hand side of Eq. (33) is represented by Line A; Line B represents the right hand side. Intersection of these two lines occurs at Point 1. The pH of this point gives the established pH, i.e. pH 9,0, which is the equivalence point for the salt of the weak acid HA.

2. Diprotic Acid Systems

A diprotic acid, H_2A, has two salts: a monovalent form - HA^- and a divalent form - $A^=$. Three equivalent solutions are identified for the system - one each for the addition of the pure acid H_2A and the pure salts HA^- and $A^=$. The equivalence points for these solutions can be determined as follows:

pH

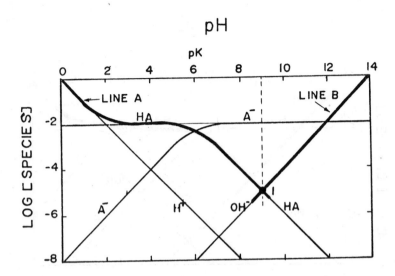

Figure 6. Addition of 10^{-2} moles/ℓ NaA to pure water (equivalent A^- solution)

(i) Addition of H_2A: 0,01 moles/litre of diprotic acid H_2A is added to water. pK values for the acid, pK_1 and pK_2 equal 6,33 and 10,33 respectively. Estimate the pH. Proton balance equation for the system is developed as follows:

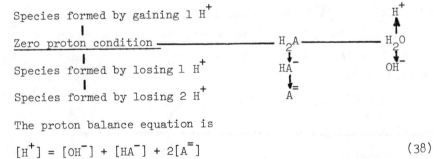

The proton balance equation is

$$[H^+] = [OH^-] + [HA^-] + 2[A^=] \tag{38}$$

The log [species] – pH diagram for the system is shown in Figure 7. Line B in this diagram represents the right hand

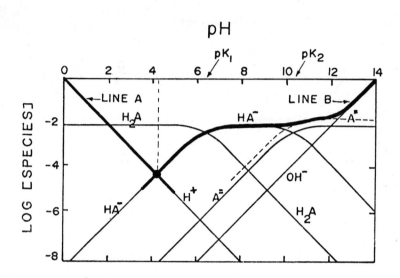

Figure 7. Addition of 10^{-2} moles/ℓ H_2A to pure water (equivalent H_2A solution)

side of the Eq. (38). Note that the term $2[A^=]$ is represented by a line parallel to the $[A^=]$ line, but log 2 (i.e. 0,303) units above it. The pH established is given by the intersection point of the $[H^+]$ line and line B, i.e. pH 4,3.

(ii) Addition of HA^-: 0,01 moles/litre of the salt of a weak diprotic acid, NaHA, is added to water, estimate the pH (pK_1 = 6,3; pK_2 = 10,3). Proton balance equation for the system is developed as follows:

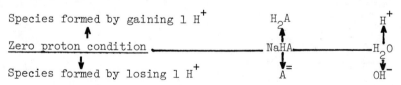

The required equation is

$$[H_2A] + [H^+] = [A^=] + [OH^-] \qquad (39)$$

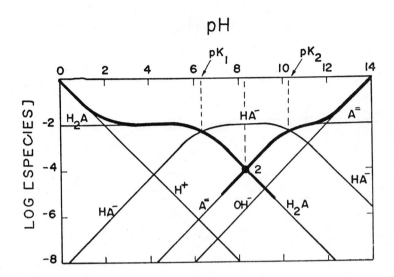

Figure 8. Addition of 10^{-2} moles/ℓ NaHA to pure water (equivalent HA^- solution)

Solution to this equation in Figure 8 gives pH 8,4 as the required equivalence point.

3. Addition of $A^=$:

0,01 moles/litre of the salt Na_2A is added to water, estimate the pH ($pK_1 = 6,3$; $pK_2 = 10,3$). Proton balance equation is

$$2[H_2A] + [HA^-] + [H^+] = [OH^-] \tag{40}$$

Solution to this equation in Figure 9 gives pH 11,5 as the equivalence point for a 0,01 M equivalent solution of the divalent salt Na_2A.

pH Established by Mixed Systems in Water

A mixed system is one that contains the species of more than one weak acid or base in solution. An example of such a system is the additon of NH_4A to water. This salt dissociates to

pH

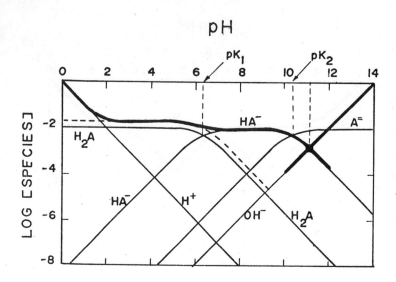

Figure 9. Addition of 10^{-2} moles/ℓ of Na_2A to pure water (equivalent $A^=$ solution)

create two weak acid systems in water: NH_4^+ - NH_3 system with pK_a 9,3 and an A^- - HA system with a pK value say 6,0. Solution procedures for estimating the pH of a variety of such systems, using the log [species] - pH diagram and a proton balance equation are given below.

1. 0,01 moles/litre of the salt NH_4A is added to water, estimate the pH. $pK_{aNH_4^+}$ 9,3; pK_{HA} 6,0. Proton balance equation for this mixed system is derived from the table set out below.

Species formed by gaining 1 H^+ HA H^+

<u>Zero proton condition</u> $\underline{}$ NH_4^+ $\longrightarrow$ A^- $\longrightarrow$ H_2O

Species formed by losing 1 H^+ NH_3 OH^-

The required equation is

$$[NH_3] + [OH^-] = [HA] + [H^+] \tag{41}$$

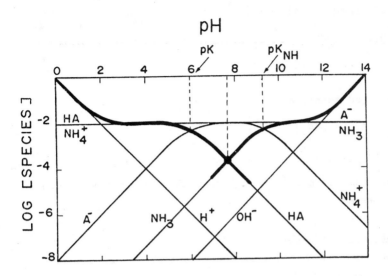

Figure 10. pH established on addition of 10^{-2} moles/ℓ of NH_4A to water water is pH 7,7

The log [species] diagram for this mixed system is shown in Figure 10. The pH for this solution is established as follows: Lines A and B represent the left and right hand sides of Eq. (41) respectively. Their intersection point defines the pH established, i.e. pH 7,7.

2. 0,01 moles/litre of HA and 0,1 moles/litre of its salt NaA are added to water, determine the pH of the mixture. pK_{HA} 6,0. To estimate the pH of this mixture it is necessary to assume that two different weak acid systems are in solution: one for the acid addition giving the $[HA_1] - [A_1^-]$ system and a second for the salt addition giving the $[HA_2] - [A_2^-]$ system. Both systems have the same pK value.

Proton balance equation is derived from the following table:

Species formed by gaining 1 H^+

Zero proton condition

Species formed by losing 1 H^+

The required equation is:

$$[A_1^-] + [OH^-] = [HA_2] + [H^+] \tag{42}$$

The log [species] diagram for these additions is shown in Figure
11. Solving for Eq. (42) in Figure 11 gives the pH for the
mixture as pH 6,9.

Alkalinity and Acidity

At this stage the meaning of equivalent solutions and equivalence points has been established in a quantitative manner. The
value of these concepts becomes evident only when mixtures of
weak acids (or their salts) with strong acids or bases are

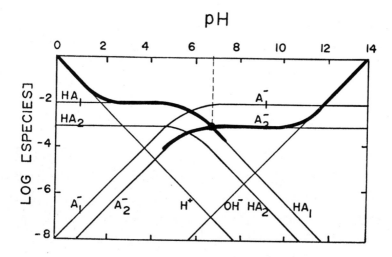

Figure 11.   pH established on addition to pure water of 10^{-2}
moles/ℓ HA and 10^{-3} moles/ℓ of its salt NaA is pH 6,9

considered. The concept of equivalent solutions allows one to separate out in a quantitative manner the molar masses of strong acids or bases and weak acids (or their salts) in a solution. In order to develop the basic ideas, consider the following examples:

(i) Determine the pH established by mixing 0,01 moles/litre of the salt NaA and 0,001 moles/litre of strong acid HCl. pK is 6,0. The proton balance equation for this mixture is established as follows:

Species formed by gaining 1 H^+ ... HA ... H^+

Zero proton condition ——————— HCl ——— NaA ——— H_2O

Species formed by losing 1 H^+ ... Cl^- ... OH^-

The required equation is

$$[HA] + [H^+] = [Cl^-] + [OH^-] \tag{43}$$

Solving this equation in the log [species] diagram for this system, Figure 12 gives pH 7,0. The molar mass of strong acid 0,001 moles/litre, is termed *the acidity of the sample.*

(ii) Determine the pH established by mixing 0,01 moles/litre of weak acid HA and 0,001 moles/litre of strong base BOH. pK_{HA} is 6,0. The proton balance equation for this mixture from the concept of proton levels is

$$[H^+] + [B^+] = [A^-] + [OH^-] \tag{44}$$

Solving this equation in the log [species] diagram Figure 13 gives pH 5,0. The strong base concentration is termed the *alkalinity of the solution.*

In the two examples above the basic concepts of Alkalinity and Acidity have been put forward. In practice of course, it is rare that C_T and the molar mass of strong acid or base will be known. The final objective is to determine from some set of

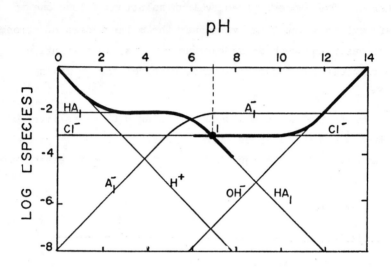

Figure 12.   pH equals 7,0 when 10^{-2} moles/ℓ of NaA and 10^{-3} moles/ℓ of HCl are added to pure water

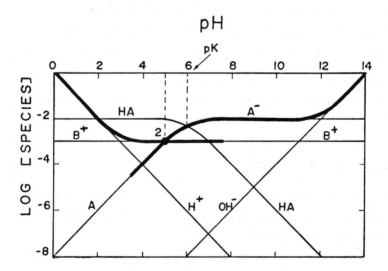

Figure 13.   pH equals 5,0 when 10^{-2} moles/ℓ of HA and 10^{-3} moles/ℓ of strong base are added to pure water

measurements C_T and the Alkalinity or Acidity and hence each of the individual species concentrations. We will approach this objective step by step. In all cases it will be found that two independent measurements on the system is all that is required for a unique solution. In the above examples C_T and the molar mass of strong acid or base were given. If C_T and pH are given a solution is also possible as illustrated in the following examples for monoprotic and diprotic weak acid systems.

A. Monoprotic Weak Acid System

A water contains a total weak acid species concentration, C_T, of 0,01 moles/litre and has a pH of 6,4. The pK for the weak acid is 6,0. This water can be interpreted as (i) an equivalent HA solution plus a net strong base or (ii) an equivalent A^- solution plus a net strong base.

(i) Proton balance equation for an equivalent HA solution plus a net strong base, BOH, is

$$[A^-] + [OH^-] = [B^+] + [H^+] \tag{45}$$

Log [species] plot for the weak acid system is shown in Figure 14. The net strong base required to give the mixture a pH of 6,4 is found by solving for $[B^+]$ in Eq. (45) using Figure 14 as follows: Line A in this diagram represents the left hand side of Eq. (45). Intersection of Line A and the vertical line for pH 6,4 occurs at Point 1. Consequently, that line representing the right hand side of Eq. (45) must also pass through Point 1, i.e. Line B giving the net strong base concentration, $[B^+]$, as 0,0063 moles/litre. This *net strong base* is known as *Alkalinity*, i.e. from Eq. (45):

$$\text{Alkalinity} = [B^+] = [A^-] + [OH^-] - [H^+] \tag{46}$$

Each of the species concentrations on the right hand side of Eq. (46) are concentrations at the observed pH.

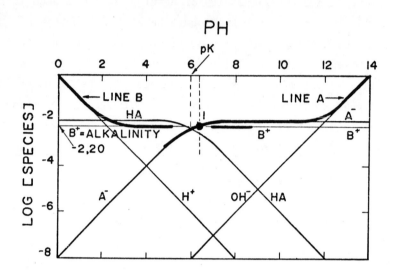

Figure 14. Measured pH = 6,4 and C_T equals 10^{-2} moles/ℓ gives Alkalinity $10^{-2,2}$ moles/ℓ

(ii) Proton balance equation for a mixture of an equivalent salt solution, A^-, and a *strong* acid, HCℓ, is:

$$[HA] + [H^+] = [C\ell^-] + [OH^-] \qquad (47)$$

Line A in Figure 15 represents the left hand side of Eq.(47). Intersection of Line A and the vertical line for pH 6, occurs at Point 1. The line representing the right hand side of Eq. (47) must also pass through this point, i.e. Line B giving the *net strong acid* concentration, equal to $C\ell^-$, as 0,0025 moles/litre. This value is known as *Acidity*, and from Eq. (47)

$$\text{Acidity} = [C\ell^-] = [HA] + [H^+] - [OH^-] \qquad (48)$$

Specific names are given to the alkalinity and acidity values associated with each equivalence point. These names are for convenience and to ensure ready identification. Unfortunately, the multiplicity of names can result in confusion and

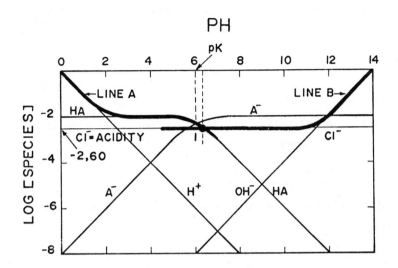

Figure 15. Measured pH = 6,4 and C_T equals 10^{-2} moles/ℓ gives Acidity $10^{-2,60}$ moles/ℓ

disguise the underlying simplicity of the system. For example, in the example above, for a weak monoprotic acid:

(1) HA equivalence point.

If the pH is *below* the HA equivalence point, the net strong acid (Acidity) superimposed on the equivalent HA solution is called *Mineral Acidity* (Eq. (49)). If the pH is *above* the HA equivalence point, the asociated net strong base (Alkalinity) is called *Total Alkalinity* or simply Alkalinity (Eq. (50)). Equations (49 and 50) for these two parameters, are derived from a proton balance for a water containing an equivalent HA solution and a net strong acid (Mineral Acidity) or a net strong base (Alkalinity) respectively.

Mineral Acidity = $[H^+] - [A^-] - [OH^-]$ (49)

Alkalinity   = $-[H^+] + [A^-] + [OH^-]$ (50)

Comparing Eq. (49 and 50), Mineral Acidity is the negative
of Alkalinity.

(2) A^- equivalence point.

If the pH is above the A^- equivalence point, the strong
base (alkalinity) superimposed on the equivalent A^- solution is
called *Caustic Alkalinity*. If the pH is below the A^- equivalence point, the associated net strong acid is called *Total
Acidity* or simply Acidity. Equations for these parameters can
be determined from a proton balance for a water containing an
equivalent A^- solution and a strong base or acid respectively,
Eqs. (51 and 52)

$$\text{Caustic Alkalinity} = [OH^-] - [H^+] - [HA] \tag{51}$$

$$\text{Acidity} = -[OH^-] + [H^+] + [HA] \tag{52}$$

Comparing Eqs. (51 and 52), the Caustic Alkalinity is the
negative of Acidity.

The expressions above give rise to four names for the various
forms of alkalinity and acidity associated with two equivalence
points but the nomemclature attached to these forms does not
indicate the specific connection with the relevant equivalence
point. A simpler and more rational terminology would be to
link the alkalinity or acidity to the particular equivalence
point, e.g.

HA Alkalinity   as against   Alkalinity

HA Acidity   as against   Mineral Acidity

A^- Alkalinity   as against   Caustic Alkalinity

A^- Acidity   as against   Acidity

The advantages and generality of this form of designation
will become even more apparent when considering diprotic acids
in the next section.

There are two further aspects which arise:

(1) For each equivalence point the alkalinity can be expressed
as negative acidity and vice versa. For example, from Eqs. (51
and 52), for a selected pH greater than the A^- equivalence
point

   A^- Alkalinity = $-A^-$ Acidity

that is, one need only to select either alkalinity or acidity
as the term to specify the condition of the weak acid-base
system. This form of expression is particularly useful in
theoretical expositions.

(2) Usually the pH is above the HA equivalence point and below
the A^- equivalence point, i.e. between the equivalence points.
Consequently, HA Alkalinity and A^- Acidity are both positive.
The custom has arisen therefore, to call the HA Alkalinity
simply 'Alkalinity' and A^- Acidity simply 'Acidity'.   Where
such usage is widespread for a particular weak acid- base system, it is perhaps best to retain that nomenclature.   In these
cases it is desirable to express the other strong acid or base
associated with the equivalence point as the negative of the
accepted parameter. For example, if Alkalinity is the accepted designation for the HA equivalence point, then if pH is
less than this equivalence point one would speak of negative
Alkalinity.

The various designations are illustrated in the following
sketches.

Forms of alkalinity for a monoprotic acid

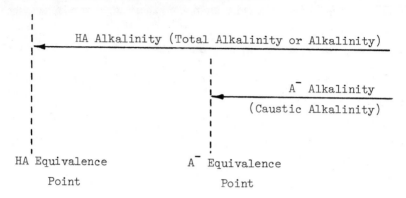

Forms of acidity for a monoprotic acid

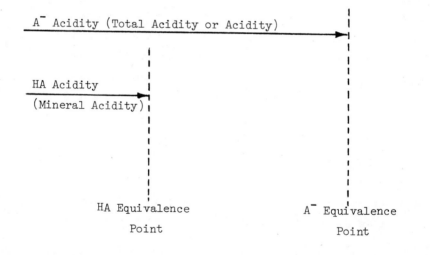

---

* Designations in brackets are those usually employed

HA Alkalinity and A^- Acidity are linked to the total species concentration C_T.   Adding Eqs.(50) and (52):

HA Alkalinity (Alkalinity) = $[A^-] + [OH^-] - [H^+]$

A^- Acidity (Acidity)   = $[HA] + [H^+] - [OH^-]$

Alkalinity + Acidity   = $[HA] + [A^-]$

$\qquad\qquad\qquad\qquad$ = C_T (53)

B. Diprotic weak acid systems

Diprotic weak acids have three equivalence points (one each for H_2A, HA^- and $A^=$) and hence three equivalent solutions. Again, for each equivalent solution there is an associated alkalinity or acidity depending on whether the pH is greater or less than the particular equivalence point respectively.

Equations for the alkalinity or acidity associated with each of the equivalence points are developed from a proton balance and are given below:

(i) H_2A equivalence point

If the pH of a water sample is above the H_2A equivalence point the associated net strong base is conventionally termed Alkalinity or Total Alkalinity in the carbonic system. Using the nomenclature developed for monoprotic acids this alkalinity is H_2A Alkalinity:

$\qquad$ H_2A Alkalinity   = $2[A^=] + [HA^-] + [OH^-] - [H^+]$ (54)
$\qquad\qquad$ (Alkalinity)

If the pH is below this equivalence point the associated net strong acid is conventionally termed Mineral Acidity. Using the new convention this acidity is H_2A Acidity, or negative H_2A Alkalinity:

$$H_2A \text{ Acidity} = -2[A^=] - [HA^-] - [OH^-] + [H^+]$$
$$\text{(Mineral Acidity)} = -H_2A \text{ Alkalinity} \qquad (55)$$

(ii) HA⁻ equivalence point

If the pH of the sample is above the HA⁻ equivalence point, the associated strong base is termed HA⁻ Alkalinity (conventionally Phenolpthalein Alkalinity for the carbonic system); if pH is below this equivalence point the net strong acid is termed HA⁻ Acidity (or negative HA⁻ Alkalinity) (conventionally the CO_2 Acidity for the carbonic system). Equations for these parameters are:

$$HA^- \text{ Alkalinity} = [A^=] + [OH^-] - [H_2A] - [H^+]$$
$$\text{(Phenolpthalein Alk.)} \qquad (56)$$

$$HA^- \text{ Acidity} = -[A^=] - [OH^-] + [H_2A] + [H^+] \qquad (57)$$
$$(CO_2 \text{ Acidity})$$

(iii) A⁼ equivalence point

If the pH is above or below this equivalence point the associated net strong base or acid is termed the A⁼ Alkalinity (conventionally Caustic Alkalinity) and A⁼ Acidity (or Acidity) respectively

$$A^= \text{ Alkalinity} = [OH^-] - [H^+] - [HA^-] - 2[H_2A] \qquad (58)$$
$$\text{(Caustic Alk.)}$$

$$A^= \text{ Acidity} = -[OH^-] + [H^+] + [HA^-] + 2[H_2A] \qquad (59)$$
$$\text{(Acidity)}$$

The various designations are illustrated in the sketches below for the carbonic system.

( ) indicates conventional nomenclature

Forms of alkalinity for a diprotic acid

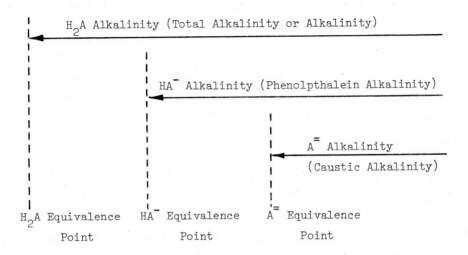

Forms of acidity for a diprotic acid

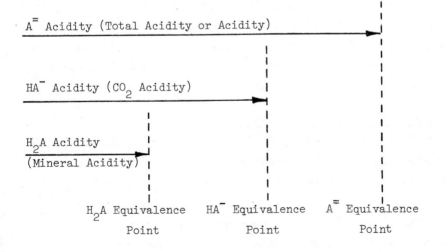

Adding Eqs. (54 and 59) gives the total weak acid species concentration, C_T, in terms of Alkalinity (H_2A Alkalinity) and Acidity ($A^=$ Acidity).

$$C_T = (Alkalinity + Acidity)/2 \tag{60}$$

Subtracting Eq. (59) from Eq. (54) gives a relationship linking Alkalinity (H_2A Alkalinity) and Acidity ($A^=$ Acidity) to either HA^- Alkalinity or HA^- Acidity.

$$HA^- \text{ Acidity} = (Acidity - Alkalinity)/2 \tag{61}$$

Because the pH of most natural waters is between the $H_2CO_3^*$ and $CO_3^=$ equivalence points, common practice is to consider only the alkalinity to the $H_2CO_3^*$ equivalence point and acidity to the $CO_3^=$ equivalence point. These two parameters are generally known by the terms Alkalinity and Acidity respectively. To avoid confusion it is therefore desirable that they be retained. In water treatment problems the pH is often adjusted outside these limits (for example, in water softening pH is adjusted to above the $CO_3^=$ equivalence point). For these cases in this report where the pH is above the $CO_3^=$ equivalence point the usual term Caustic Alkalinity will not be used but the term negative Acidity; where the pH is below the $H_2CO_3^*$ equivalence point the usual term Mineral Acidity will not be used but the term negative Alkalinity. These modifications are necessary in order to avoid complexity in the quantitative theoretical development of the system.

Alkalinity and Acidity Measurement

It has been shown (see page 22) that if values for the two parameters C_T and pH are known, all the parameters for the weak acid system can be determined. However, in practice C_T is often unknown and cannot be directly measured. Alkalinity (or Acidity) and pH, however, usually can be measured. If values of any two of these three parameters are known then sufficient information is available to solve for the complete system.

The determination of the equivalence points of a sample may present a number of problems: (i) the equivalence point may change with C_T (which itself is initially unknown); (ii) if in the pH region of the equivalence point the buffer capacity is large, then significant titration errors are inherent in the determination; (iii) if the pK value of the weak acid system is too high or too low (i.e. far removed from pH 7) the buffer action of the water species masks the manifestation of the equivalence point.

These problems make it necessary to investigate thoroughly the characteristics of the weak acid system in order to select the most useful equivalence points and obtain the maximum accuracy in determining a particular alkalinity or acidity.

Identification of equivalence points by plotting a 'titration curve' is important and leads to alternative procedures for determining the end point for certain of the weak acid systems particularly the carbonic system.

If a strong acid or base is incrementally added to a sample, and pH is monitored after each addition, a plot of the total molar mass of strong acid (or base) added versus pH gives the alkalimetric (or acidimetric) titration curves (see Figure 16).

A theoretical plot of the alkalimetric or acidimetric titration curves can be derived for a particular system using the log[species]-pH diagram and a proton balance equation for addition of a strong acid or base to a weak acid system.   For example, 0,01 moles/ℓ of a weak acid HA (pK_a = 5,1) is added to a water.   A titrant of strong base, BOH, is then incrementally added.

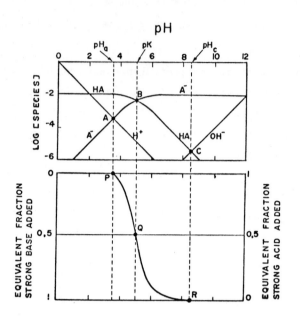

Figure 16. Log [species] - pH diagram and alkalimetric (or acidimetric) titration curve for monoprotic weak acid system

pH before titrant is added: The pH established is given by the
equivalence point for 0,01 M equivalent HA solution, pH 4,0 in
Figure 17.

pH after each incremental addition of strong base: The pH established after each incremental addition of strong base is
obtained by solving the relevant proton balance equation in the
log [species] - pH diagram. Proton balance equation for addition of strong base to an equivalent weak acid solution is

$$[B^+] + [H^+] = [A^-] + [OH^-] \tag{62}$$

As the titration progresses, the horizontal line representing each new value for the total molar mass of titrant added,
$[B^+]$ in Figure 17, is drawn on the log [species] - pH diagram.
Equation (62) is then solved graphically for each $[B^+]$ to give
corresponding pH values. A plot of pH against titrant added
gives the theoretical titration curve shown in Figure 17. This

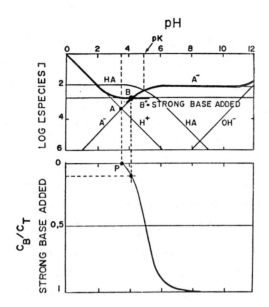

Figure 17. Graphical development of titration curve for weak
acid system $C_T = 10^{-2}$ moles/ℓ

curve allows determination of the equivalence points.

Determination of Equivalence Point by Titration

The shape of the titration curve yields information regarding
the properties of the weak acid sample being titrated: the pH
values of specific inflection points in the curve correspond
to the pK value(s) and acid and salt equivalence points for
the weak acid.   Three inflection points are observed in the
titration curve for a monoprotic weak acid system (Figure 16).
Comparing the log [species] - pH diagram and titration curve
for this system, Figure 16, the following observations are per-
tinent:

(i) The pH of the inflection points P and R in the
        titration curve correspond to the equivalence
        points A and C respectively in the log [species]
        - pH diagram; these equivalence points refer to
        the equivalent solutions HA and A^- respectively.

(ii)   The pH of the inflection point Q in the titration
        curve corresponds to pH equal to pK_a in the log
        [species]-pH diagram. If a strong base is added
        to an equivalent HA solution (pH given by Point P)
        until pH reaches the next equivalence point, Point
        R for an equivalent A^- solution, the base to be
        added to give pH = pK_a (where pK_a is between pH_P
        and pH_R) equals half that required to change the
        pH from P to R. This is also true for titrations
        between consecutive equivalence points for poly-
        protic weak-acid-base systems.

(iii) Inspection of the titration curve in Figure 16
        shows a continuous variation in the change of pH
        for unit additions of titrant over the range
        shown. The maximum slope in the titration curve
        occurs at the pH value of pK_a, i.e. where the
        concentrations of weak acid, HA, and its salt, A^-,
        are equal. At this point the smallest change in
        pH occurs for a given addition of titrant. The
        minimum slope in the titration curve occurs where
        the pH values are for the equivalent HA and A^-
        solutions, i.e. at these pH values the largest
        change in pH occurs for a given addition of ti-
        trant. The slope of the titration curve intro-
        duces the concept of 'buffer action' for weak
        acids and bases. This parameter is very useful
        for practical determination of the accuracy of
        alkalinity and acidity measurements by titration
        to the respective inflection points.

Buffer Index

A measure of the buffer capacity is the concentration of strong
acid or base required to change the pH by a given amount. The

larger the mass of strong acid or base to be added to do so, the better the buffer.

Buffer index (or buffer capacity),β, is defined for the addition of δC_B moles/ℓ strong base as (van Slijke, D.D., 1922):

$$\beta = \frac{\delta C_B}{\delta pH} \qquad (63)$$

Addition of a strong acid has the reverse effect on pH to the addition of a strong base. β can be written also in terms of the strong acid added (δC_A) as:

$$\beta = - \frac{\delta C_A}{\delta pH}$$

β is a positive number. Thus, buffer index at some pH is the slope of the titration curve at that pH, i.e. the concentration of strong base or acid required to give a unit change in pH.

The buffer index equation for a monoprotic weak acid and its salt has a simple form. (Butler, J.N., 1964). Consider a solution containing C_T moles/ℓ of a weak acid and its species and C_B moles/ℓ of a strong base, NaOH. Equilibria equations for the solution are:

$$\frac{[H^+] [A^-]}{[HA]} = K_a \qquad (64)$$

and

$$[H^+][OH^-] = K_w \qquad (65)$$

Mass balance equations are:

(i) for the mass of weak acid and its species in solution:

$$[HA] + [A^-] = C_T \qquad (66)$$

(ii) for the mass of strong base, NaOH, in solution:

$$[Na^+] = C_B \qquad (67)$$

For electro-neutrality, a charge balance gives the following
equation:

$$[H^+] + [Na^+] = [OH^-] + [A^-] \tag{68}$$

Using Eqs.(62) to (68) an equation can be derived for buffer
capacity, β, in terms of the hydrogen ion concentration, H^+,
and the total mass of weak acid species in solution C_T. Solving
for [HA] in Eq. (66), substituting into Eq. (64) and solving
for [A$^-$]:

$$[A^-] = \frac{C_T \cdot K_a}{K_a + [H^+]} \tag{69}$$

Substituting Eqs. (67) and (69) into Eq. (68):

$$[H^+] + C_B = K_w/[H^+] + C_T K_a/(K_a + [H^+])$$

i.e.

$$C_B = K_w/[H^+] - [H^+] + C_T K_a/(K_a + [H^+]) \tag{70}$$

Differentiating Eq. (70) with respect to pH gives an equation
for buffer capacity, β, as defined by Eq. (62), i.e.

$$\beta = \frac{\delta C_B}{\delta pH}$$

$$= \frac{\delta C_B}{\delta [H^+]} \cdot \frac{[H^+]}{\delta pH} \tag{71}$$

Differentiating Eq. (70) with respect to [H$^+$] gives:

$$\frac{\delta C_B}{\delta [H^+]} = \frac{-K_w}{[H^+]^2} - 1 - \frac{C_T K_a}{(K_a + [H^+])^2} \tag{72}$$

The term $\delta[H^+]/\delta pH$ is solved for as follows:

$$pH = -\log_{10}[H^+]$$

Rewriting the above equation using the natural logarithm of $[H^+]$:

$$pH = \frac{-1}{2,303} \cdot \ln [H^+]$$

i.e.

$$\frac{\delta pH}{\delta [H^+]} = -1/(2,303 \cdot [H^+])$$

i.e.

$$\delta [H^+]/\delta pH = -2,303 \cdot [H^+] \tag{73}$$

Substituting Eqs. (72) and (73) into Eq. (71):

$$\frac{\delta C_B}{\delta pH} = (\frac{-K_w}{[H^+]^2} - 1 - \frac{C_T K_a}{(K_a + [H^+])^2}) \cdot (-2,303 \cdot [H^+])$$

i.e.

$$\frac{\delta C_B}{\delta pH} = 2,303 (\frac{K_w}{[H^+]} + [H^+] + \frac{C_T K_a [H^+]}{(K_a + [H^+])^2})$$

i.e.

$$\beta = 2,303 (\frac{K_w}{[H^+]} + [H^+] + \frac{C_T K_a [H^+]}{(K_a + [H^+])^2}) \tag{74}$$

The first two terms of Eq. (74) result from the buffering effect of water and the third term from the buffering effect of the weak acid. Eq. (74) is thus rewritten as

$$\beta = \beta_{H_2O} + \beta_{Ha}$$

where

$$\beta_{H_2O} = 2,303 (K_w/[H^+] + [H^+])$$

and

$$\beta_{HA} = \frac{2,303 \cdot K_a \cdot C_T \cdot [H^+]}{(K_a + [H^+])^2}$$

Figure 18 shows the distribution of buffer capacity with pH
for a 0,01 M solution of acetic acid pK_a = 5,1. Examining the
buffer capacity curves shows that this weak acid solution af-
fords maximum buffer capacity at a pH where $pH = pK_a$, regions
of minimum buffer capacity occurring on both sides of this maximum. The bell-shaped curve for β_{HA} in Figure 18 always has the
same shape for the same C_T value irrespective of the type of
weak acid or base used as the buffer, i.e. irrespective of the
value of K_a. For any weak acid, HA, with some particular K_a
value the curve β_{HA} for unit C_T may be traced from a master
curve by merely shifting the pH axis so that the maximum occurs
at $pH = pK_a$. The unit curve values are multiplied by C_T to
give β_{HA} curve.

A different form of the equation for buffer capacity, β, in
terms of the weak acid species in solution can be derived by
substituting for K_a and C_T from Eqs. (64) and (66) into Eq.(74)
and simplifying:

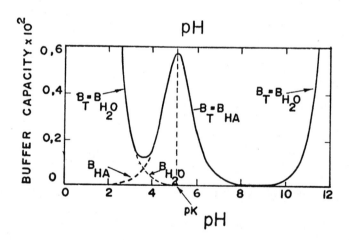

Figure 18. Distribution of buffer capacity with pH for a 10^{-2}M
weak acid system in water

$$\beta = 2,303 \left(\frac{K_w}{[H^+]} + [H^+] + \frac{[HA]\,[A^-]}{[HA] + [A^-]} \right) \tag{75}$$

Buffer Index of a Mixture of Monoprotic Acids

The buffer index, β, of a mixture of monoprotic acids and bases
has a simple dependence on pH. The expression derived for β of
a solution containing one acid-salt pair, Eq. (74), is easily
generalized for any number of acid-salt pairs.

If a charge balance for a mixture of several weak acids, HA_1,
HA_2, HA_3, etc., is expressed in terms of $[H^+]$ then using the
various equilibria expressions and differentiating with respect
to pH, an expression for the buffer index is derived containing
a series of additive terms - each pertaining to one component of
the mixture. Thus

$$\beta = \beta_{H_2O} + \beta_{HA_1} + \beta_{HA_2} + \ldots\ldots \tag{76}$$

where
$$\beta_{H_2O} = 2,303\,([H^+] + K_w/[H^+])$$

and
$$\beta_{HA_x} = 2,303\,\frac{C_T \cdot K_{ax} \cdot [H^+]}{(K_{ax} + [H^+])^2}$$

and
$$x = 1,2,3 \ldots\ldots$$

Figure 19 shows the total buffer index diagram and the con-
stituent buffer curves for a solution containing a number of
monoprotic acids. Assuming that the total species concentration, C_{Tx}, is the same for all the weak acids, the bell-shaped
buffer index curve for each acid has the same shape but in each
case the curve is plotted about the particular acid's pKa value.

Buffer Index of a Diprotic Acid

An exact expression for buffer index of C_T moles/ℓ of a diprotic acid H_2A, is derived by a similar method as the expression

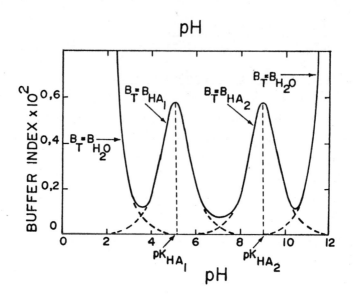

Figure 19. Distribution of buffer capacity with pH for a water containing equal concentration of two monoprotic weak acid systems

for β of a monoprotic acid (Eq. 74). This yields β for a diprotic acid as:

$$\beta = \beta_{H_2O} + \beta_{H_2A} \tag{77}$$

where

$$\beta_{H_2A} = 2{,}303 \cdot C_T \cdot K_1 \cdot [H^+] \left(\frac{[H^+]^2 + 4K_2[H^+] + K_1K_2}{([H^+]^2 + K_1[H^+] + K_1K_2)^2} \right) \tag{78}$$

Butler (1962) showed that where the ratio of successive ionization constants of a polyprotic acid is less than 5 per cent, the polyprotic acid may be considered as a mixture of monoprotic acids of equal concentrations. Buffer index β is then approximated by Eq. (76).

Suppose a solution contains C_T moles/ℓ of diprotic acid H_2A. Assuming H_2A can be considered as a mixture of two monoprotic acids each with total species concentrations of C_T moles/ yields the approximate expression of β,

$$\beta = \beta_{H_2O} + \beta_{H_2A} + \beta_{HA}$$

i.e.

$$\beta = 2{,}303([H^+] + \frac{K_w}{[H^+]}) + \frac{2{,}303\ C_T[H^+]K_1}{(K_1 + [H^+])^2} + \frac{2{,}303\ C_T\ K_2[H^+]}{(K_2 + [H^+])^2} \quad (79)$$

Rewriting the final two terms of Eq. (79) in the same form as Eq. (79) yields

$$\beta_{H_2A} + \beta_{HA} = 2{,}303\ C_T K_1 [H^+] \quad (80)$$

$$(\frac{[H^+]^2(1+K_2/K_1) + 4K_2[H^+] + K_1K_2(1+K_2/K_1)}{([H^+] + K_1[H^+](1+K_2/K_1) + K_1K_2)^2}$$

Comparing exact equation for β, Eq. (78), with approximate equation, Eq. (80), shows that approximate solution contains an extra factor $(1 + K_2/K_1)$. Provided K_2 is less than 5 per cent of K_1, this factor introduces an error less than 5 per cent in the value of β.

Buffer Capacity, Log [species] - pH Diagram and Titration Curves

In Figure 20 the log [species] - pH diagram, buffer capacity diagram and titration curve are shown for a 10^{-2} monoprotic weak acid system, $pK_a = 6{,}0$. Comparing these three diagrams the following points are noted:

(a) pH of maximum and minimum buffer capacity correspond to inflection points in the titration curve.

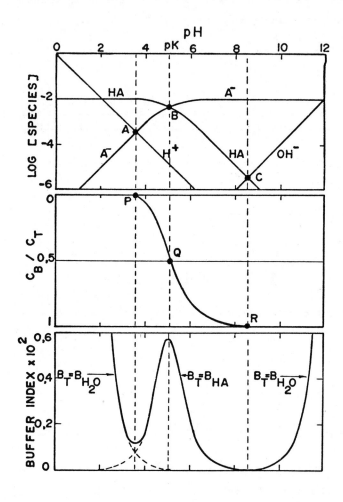

Figure 20. Log [species] - pH diagram, titration curve and buffer capacity diagram for a weak monoprotic acid $C_T = 10^{-2}$ moles/ℓ

(b) points of minimum buffer capacity correspond to pH values for an equivalent HA solution, pH_A, and an equivalent A^- solution, pH_C.

(c) The point of maximum buffer capacity corresponds to the titration midpoint where [HA] equals [A^-] and pH equals pK_a.

These observations can be deduced from basic theory as follows:

(1) HA equivalence point:

The pH of an equivalent HA solution corresponds to an inflection point in the titration curve and a point of minimum buffer capacity.

From the definition of buffer capacity, $\beta = -\delta C_A/\delta pH$, the pH value of a minimum in buffer capacity corresponds to an inflection point in the titration curve; that this pH also corresponds to an equivalence point is shown as follows:

Proton balance equation for addition of weak acid HA to pure water is:

$$[H^+] = [OH^-] + [A^-] \tag{81}$$

For the point of minimum buffer capacity in this region $\delta\beta/\delta pH$ equals zero. Differentiating Eq. (74) with respect to pH and equating to zero:

$$\frac{\delta\beta}{\delta pH} = 2,303^2([OH^-] - [H^+] - \frac{C_T K_a [H^+]}{(K_a + [H^+])^2} + \frac{2C_T K_a [H^+]^2}{(K_a + [H^+])^3}) = 0 \tag{82}$$

Substituting for $C_T K_a [H^+]/(K_a + [H^+])^2$ from Eq. (75) into the above equation and simplifying

$$[OH^-] - [H^+] - \frac{[HA][A^-]}{[HA]+[A^-]} + \frac{2[HA]^2[A^-]}{([HA]+[A^-])^2} = 0 \tag{83}$$

At the HA equivalence point pH $\leftarrow$ pK_a, [HA] $\rightarrow$ [A$^-$] and Eq.
(83) simplifies to

$$[OH^-] - [H^+] + [A^-] = 0$$

which is the same equation as the proton balance for addition
of HA to water, Eq. (81). Thus, the pH value for the HA equivalence point corresponds to an inflection point in the titration curve.

(2) A$^-$ equivalence point:

Proton balance equation for addition of the salt A$^-$ to pure
water is

$$[OH^-] = [HA] + [H^+] \qquad\qquad (84)$$

For the point of minimum buffer capacity in this pH region:
pH $\rightarrow$ pK_a, [A$^-$] $\rightarrow$ [HA] and $\delta\beta/\delta$pH equals zero. Equation (83)
approximates to

$$[OH^-] - [H^+] - [HA] = 0$$

This is the same equation as that for addition of salt A$^-$
to water, Eq. (84). Thus the pH of the inflection point in
the titration curve corresponds to the A$^-$ equivalence point.

(3) pH equals pK$_a$ i.e.[HA] = [A$^-$]

If a point of maximum buffer capacity occurs at pH equal to
pK_a, then $\delta\beta/\delta$pH must equal zero at this pH. Substituting [HA]
equal to [A$^-$] in Eq. (83)

$$\frac{\delta\beta}{\delta pH} = 2{,}303^2([H^+] - [OH^-])$$

If [H$^+$] and [OH$^-$] are both negligible at this pH, then
$\delta\beta/\delta$pH closely approximates to zero. Thus pH equal to pK_a
corresponds closely to a pH of maximum buffer capacity and an
inflection point in the titration curve. If however, either
[H$^+$] or [OH$^-$] are not negligible, no inflection point will be

observed at pH = pK_a.

Non-occurrence of Inflection Points

The analyses above assumed that pK_a was either very much less
or much greater than the pH equivalence points. If the HA
equivalence point pH is not much less than pK_a then the basis
for the analysis is no longer valid. In this event a minimum
does not occur in the buffer capacity at this equivalence point
and no inflection point develops in the titration curve. This
situation is illustrated in the log[species]-pH diagram in
Figure 21a.

Similarly, if the A^- equivalence point pH is not much
greater than pK_a then again no minimum occurs in the buffer
capacity at this equivalence point and no inflection point in
the titration curve. The situation is illustrated in Figure 21b.

Considering the HA equivalence point, an inflection in the
titration curve will just become manifest when the pH of the
equivalence point equals pK_a for the weak acid species (see
Figure 22).

Precision of Alkalinity and Acidity Measurements

Each of the forms of alkalinity and acidity can be measured by
titrating a water sample to a particular equivalence point. The
equivalence point may be selected by plotting the titration
curve and identifying the corresponding inflection point in
this curve, or it may be predetermined using some other experimental method (to be discussed later in this monograph). In
either case an uncertainty exists as to the exact value of this
equivalence point. The cause of uncertainty may arise principally from two factors:

(1) the pH meter only reads to a certain significant figure
and consequently exact pinpointing of the inflection point is
impossible.

(2) Depending on the buffer capacity at the titration endpoint, small errors in the estimation of equivalence points may lead to large or small errors in alkalinity or acidity measurements.

The error resuting from a titration to an incorrect end-point can be determined from the theory of buffer capacity as follows.   From the definitions of buffer capacity and alkali-metric and acidimetric titrations

$$\frac{\delta[\text{alkalinity}]}{\delta pH} = \frac{-\delta[\text{acidity}]}{\delta pH} = \beta$$

Rewriting the above equation in the difference form

$$\frac{\Delta[\text{alkalinity}]}{\Delta pH} = \frac{-\Delta[\text{acidity}]}{\Delta pH} = \beta \tag{85}$$

and from Eq. (75), β for a monoprotic acid is

$$\beta = 2,303\ ([H^+] + [OH^-] + \frac{[HA]\ [A^-]}{[HA] + [A^-]}) \tag{86}$$

(i) <u>Error associated with the HA equivalence point</u>

In the pH region of the HA equivalence point $[HA] \rightarrow\!\!\!\rightarrow [A^-]$ and $[H^+] \rightarrow\!\!\!\rightarrow [OH^-]$. Eq. (86) approximates to

$$\beta = 2,303([H^+] + [A^-]) \tag{87}$$

At the equivalence point pH_e (i.e. $[H^+]_e$), $[H^+]_e \approx [A^-]$ and the equation for buffer capacity, Eq. (86), approximates to

$$\beta = 4,606.[H^+]_e$$

The error for an incorrect titration to this equivalence point can now be determined from Eq. (85)

$$\Delta[\text{Total Alkalinity}] = 4,606\ [H^+]_e.\Delta pH$$

$$\Delta[\text{Mineral Acidity}]\ \ = 4,606\ [H^+]_e.\Delta pH$$

Where ΔpH is the uncertainty in the equivalence point.

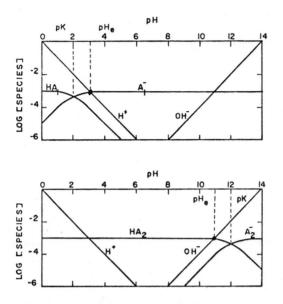

Figure 21.   Marking of inflection points by water species for relatively strong acids and bases

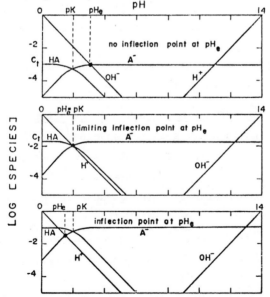

Figure 22.   Effect of C_T on manifestation of inflection point for relatively strong weak-acid systems

(ii) <u>Error associated with the A^- equivalence point</u>

At the A^- equivalence point $[A^-] \twoheadrightarrow [HA]$, $[OH^-] \twoheadrightarrow [H^+]$ and $[HA] \approx [OH^-]$. Eq. (86) approximates to

$$\beta = 4,606[OH^-]_e$$

where $[OH^-]_e$ is the hydroxyl ion concentration at the A^- equivalence point. Thus, from Eq. (85)

$$\Delta[\text{Caustic Alkalinity}] = 4,606.[OH^-]_e.\Delta pH$$

$$\Delta[\text{Total Acidity}] = 4,606[OH^-]_e.\Delta pH$$

Note that the above error analyses have great practical importance in chemical analysis of waters with low Alkalinity or Acidity. For example, analysis of the waters of the Western Cape give pH 5,8 and Alkalinity 2ppm as $CaCO_3$. Calculation of Acidity from these measurements gives Acidity = 20 ppm as $CaCO_3$. If the actual error in Alkalinity measurement is only 1 ppm as $CaCO_3$, the resulting error in the calculated value for Acidity is 10 ppm as $CaCO_3$. For waters with low Alkalinity or Acidity values measurement of these parameters by titration to an inflection point is impractical, a more reliable method based only on pH measurement is proposed in Chapter 5.

<u>Effect of Ionic Concentration on Chemical Equilibria</u>

All equilibria equations given so far have assumed ideal solutions. However, in a solution containing ions the ionic interaction between solute particles extends over a much greater distance than in a solution containing uncharged molecules. This results in deviations from ideal laws used in developing equilibria equations. Ideal laws assume each particle in solution behaves independently of any other particle, and thus hold only for very dilute ionic solutions. The long range effect of ions in solution is electrostatic attraction between oppositely charged particles. This attraction causes the activity of an ion to be smaller than its concentration. Because deviation from ideal behaviour results from electrostatic attraction, the degree of deviation of ions will depend on their charge.

Ionic Strength

An empirical measure of all ions in solution giving rise to deviation from ideal behaviour for chemical equilibria is the ionic strength.

$$\text{Ionic strength} = \mu = 1/2 \; \Sigma \; C_i \; Z_i^2 \tag{88}$$

C_i = molar concentrations of ith ion in solution

Z_i = ionic charge of the ith ion in solution

Activity Coefficients

Assuming that ions are point charges in a continuous medium of dielectric constant equal to that of water, Debye and Hückel derived a theoretical form for active concentrations of ions in sufficiently dilute solution. They showed that to apply the laws of mass action to ionic equilibrium reactions, the molar concentrations of reacting ions should be reduced by a factor known as the *activity coefficient*. The reduced ionic concentration is called the *active ionic concentration*, i.e.

$$f_i[X] = (X)$$

where

f_i = activity coefficient

[X] = molar concentration of ion X

(X) = active concentration of ion X

Thus, if a weak acid HA is added to pure water, equilibria reactions for the solution are

$$HA \rightleftarrows H^+ + A^- \quad \text{and}$$

$$H_2O \rightleftarrows H^+ + OH^-$$

Equilibria equations for these reactions are

$$(H^+)(A^-)/(HA) = K_a \quad \text{and} \tag{89}$$

$$(H^+)(OH^-) \quad = K_w \tag{90}$$

where ( ) indicate active concentrations and K values are the active equilibria constants.

Eqs. (89) and (90) can now be rewritten with species concentrations in molar form:

$$f_{H^+}[H^+].f_{A^-}[A^-]/[HA] = K_a \quad \text{and} \tag{91}$$

$$f_{H^+}[H^+].f_{OH}[OH^-] \quad = K_w \tag{92}$$

[HA] is unchanged and as such its activity is unity.

It is often convenient in water chemistry to have the left-hand side of equilibria equations only in terms of the molar concentrations of reacting species. For this purpose activity coefficients are transferred to the righthand side of the equation and incorporated in the equilibrium constant, K, to give an effective equilibrium constant, K', i.e.

From equations (91) and (92)

$$[H^+][A^-]/[HA] = K_a/f_{H^+} \cdot f_{A^-} = K'_a$$

and

$$[H^+][OH^.] \quad = K_w/f_{H^+} \cdot f_{OH^-} = K'_w$$

The last equation must be used with caution when $[H^+]$ is determined by pH measurement as pH is the negative log of the active hydrogen ion concentration. The problem is thoroughly discussed in Chapter 3.

Many expressions have been proposed for calculating activity coefficients of ions in solution. The extended Debye-Hückel law proposes that the activity coefficient, f_i, of a single ion, i, of charge Z_i, is related to the ionic strength, μ, as follows

$$\log f_i = -A \cdot Z_i^2 \cdot \frac{\sqrt{\mu}}{1 + B \cdot a \cdot \sqrt{\mu}} \qquad (93)$$

where

Z = ionic charge

A = constant which depends on absolute temperature and
dielectric constant ε of the water

$A = 1{,}825 \cdot 10^6 \cdot (\varepsilon T)^{3/2}$

$\varepsilon = 78{,}30$ for water

B = constant which is a function of the absolute temperature (T) and the dielectric constant of the
water, ε

$B = 50{,}3 \, (\varepsilon T)^{-1/2} = 0{,}328$ at 25°C in water

a = an adjustable parameter which corresponds roughly
to the effective size of the hydrated form of the
ion in question (measured in Angström units
10^{-10}M)

In water chemistry an approximation often used to the extended Debye-Hückel equation, Eq. (93), for water with an ionic
strength of less than 0,05 is the Guntelberg approximation:

$$\log f_i = -0{,}5 \cdot \frac{Z_i \cdot \sqrt{\mu}}{1 + \sqrt{\mu}} \quad \text{at } 25°C \qquad (94)$$

Guggenheim suggested that a fixed value of 3,0 A° be taken
for 'a' in the extended Debye-Hückel equation, Eq. (93) and

that data be fitted by addition of a linear term:

$$\log f_i = \frac{-A \cdot z_i^2 \cdot \sqrt{\mu}}{1 + \sqrt{\mu}} + b\mu \qquad (95)$$

Upon examination of values 'b' used in Guggenheim's equation above for a number of electrolytes, Davies proposed the following equation for calculating the activity coefficient:

$$\log f_i = -0,5 \cdot z^2 \left(\frac{\sqrt{\mu}}{1 + \sqrt{\mu}} - 0,2\mu \right) \qquad (96)$$

Experimental comparison of the activity coefficients predicted by the Guntelberg and Davies equations, Eqs. (94) and (96) has been carried out for a number of acids. The Guntelberg equation falls below experimental results, but the values predicted by the Davies equation gives a good estimate of the activity coefficients. For ionic strengths of 0,1 and 0,5 the error in prediction using the Davies equation is less than 3 per cent and 8 per cent respectively.

Example

A weak diprotic acid (dissociation constants K_1 and K_2) is added to water. Analysis of the ions present in the water yields:

$$SO_4^= \quad = \quad 20 \ mg/\ell$$

$$Cl^- \quad = 120 \ mg/\ell$$

$$Ca^{++} \quad = \quad 60 \ mg/\ell$$

$$Mg^{++} \quad = \quad 20 \ mg/\ell$$

(i) Calculate the ionic strength of the water.

(ii) Calculate the activity coefficients for univalent and divalent ions in the water at 25°C using:

(a) Guntelberg approximation to the extended Debye-Hückel law, i.e. Eq. (94)

(b) Davies equation, Eq. (96)

(iii) Write down the two equilibrium equations for the dissociation of a diprotic weak acid, H_2A, in the water in terms of:

(a) active concentrations,

(b) molar concentrations

(i) Ionic strength = $\mu = 0,5 \ \Sigma_i \ c_i \ z_i^2$; i.e.

Ion	z_i	c_i moles/ℓ	$c_i \ z_i^2$	$0,5 \ c_i \ z_i^2$
$SO_4^=$	2	$0,208 \times 10^{-3}$	$0,83 \times 10^{-3}$	$0,42 \times 10^{-3}$
Cl^-	1	$3,33 \times 10^{-3}$	$3,33 \times 10^{-3}$	$1,67 \times 10^{-3}$
Ca^{++}	2	$1,50 \times 10^{-3}$	$6,00 \times 10^{-3}$	$3,00 \times 10^{-3}$
Mg^{++}	2	$0,834 \times 10^{-3}$	$3,34 \times 10^{-3}$	$1,67 \times 10^{-3}$

i.e. $0,5 \ \Sigma_i \ c_i \ z_i^2 = (0,42 + 1,67 + 3,00 + 1,67) \times 10^{-3}$

i.e. $= 6,76 \times 10^{-3}$

(ii) (a) Guntelberg approximation

$$\log f_i = \frac{-0,5 \ . \ z_i^2 \ \sqrt{\mu}}{1 + \sqrt{\mu}} \quad \text{at } 25°C$$

for monovalent ions (i.e. activity coefficient f_m):

$$\log f_m = \frac{-0,5 \ (1)^2 \ . \ \sqrt{6,76} \ . \ 10^{-3}}{1 + \sqrt{6,76} \ . \ 10^{-3}} = -0,038$$

i.e. $f_m = 0,917$

for divalent ions (activity coefficient f_d):

$$\log f_d = \frac{-0,5 \ . \ (2)^2 \ . \ \sqrt{6,76} \ . \ 10^{-3}}{1 + \sqrt{6,75} \ . \ 10^{-3}}$$

i.e. $f_d = 0,706$

(b) The Davies equation is:

$$\log f_i = 0,5 \ . \ z^2 (\frac{\sqrt{\mu}}{1 + \sqrt{\mu}} - 0,2 \ \mu)$$

for monovalent ions

$$-\log f_m = 0,5 \ . \ z^2 (\frac{\sqrt{6,76} \ . \ 10^{-3}}{1 + \sqrt{6,76} \ . \ 1)^{-3}} - 0,2 \ . \ 6,76 \ . \ 10^{-3})$$

$$= 0,5 \ . \ 1 \ (0,0756 - 0,0013)$$

i.e. $f_m = 0,917$

for divalent ions

$$-\log f_d = 0,5 \ . \ (2)^2 \ (0,0743)$$

$$= 0,1486$$

i.e. $f_d = 0,71$

(iii) Dissociation reactions for the diprotic weak acid H_2A
in water are:

$$H_2A \rightleftarrows H^+ + HA^-$$

and $HA^- \rightleftarrows H^+ + A^=$

and $H_2O \rightleftarrows H^+ + OH^-$

   (a) Equilibrium equations involving active concentrations
of H^+, HA^-, $A^=$ and OH^- are:

For the dissociation of H_2A

$$(H^+)(HA^-)/(H_2A) = K_1 \tag{97}$$

i.e. $(f_m \cdot [H^+]) \cdot (f_m \cdot [HA])/[H_2A] = K_1 \tag{98}$

For the dissociation of HA^-

$$(H^+)(A^=)/(HA^-) = K_2 \tag{99}$$

i.e. $(f_m \cdot [H^+]) \cdot (f_d \cdot [A^=])/(f_m \cdot [HA^-]) = K_2 \tag{100}$

For the dissociation of water

$$(H^+)(OH^-) = K_w \tag{101}$$

i.e. $(f_m \cdot [H^+]) \cdot (f_m \cdot [OH^-]) = K_w \tag{102}$

where ( ) indicate active concentrations and [ ] molar
concentrations.

   (b) Equilibrium equations for the dissociation of
H_2A can be written in terms of molar concentrations.

From Eq. (98):

$$[H^+][HA^-]/[H_2A] = K_1/f_m^2 = K'_1$$

From Eq. (100):

$$[H^+][A^=]/[HA^-] = K_2/f_d = K'_2$$

and from Eq. (102):

$$[H^+][OH^-] = K_w/f_m^2 = K'_w$$

References

Bard,A.J. 1966 *Chemical Equilibrium*, Harper and Row, N.Y.

Bjerrum,N. 1914 "The Theory of Alkalimetric and Acidimetric Titration", Saamlung Chem. Techn. Vortrage, 21, 69.

Butler,J.N. *Ionic Equilibrium: A Mathematical Approach*, Addison-Wesley, Reading, Mass.

Ham,R.K. 1969 "Application of Some Fundamental Equilibrium Concepts to Water Systems", W. & S.A., 90-104.

Hamer,P., Jackson, J. & Thurston,E.F. 1961 *Industrial Water Treatment Practice*, Butterworths, London.

Kleijn,H.F.W. 1965 "Buffer Capacity in Water Chemistry" J.Air Wat.Poll., 9, 401-413.

Sillen,L.G. 1959 *Graphic Presentation of Equilibrium Data, Treatise on Analytical Chemistry*, Interscience Publ.Inc., N.Y.

van Slijke,D.D. 1922 "On Measurement of Buffer Values", J.Biol.Chem., 52, 525-590.

Weber,W.J. and Stumm,W. 1963 "Mechanism of Hydrogen Ion Buffering in Natural Waters, J.A.W.W.A., 55 1553-1578,Dec.

Chapter 3

IONIC EQUILIBRIA OF THE CARBONIC SPECIES IN WATER

Carbon dioxide dissolved into water exists not only as dissolved CO_2 but also as carbonic acid, H_2CO_3, and the ions HCO_3^- and $CO_3^=$. The sum of these concentrations in solution is the total carbonic species concentration.

The carbonic species together with the OH^- and H^+ ions of the water exist in a state of dynamic equilibrium described by the following reactions:

$$CO_2 + H_2O \rightleftarrows H_2CO_3 \tag{1}$$

$$H_2CO_3 \rightleftarrows H^+ + HCO_3^- \tag{2}$$

$$HCO_3^- \rightleftarrows H^+ + CO_3^= \tag{3}$$

$$H_2O \rightleftarrows H^+ + OH^- \tag{4}$$

The concentrations of each species are described by a series of dissociation equations.

The dissociation constant for the carbonic acid in reaction (2) is usually referred to as the "real first dissociation constant, K_r"

$$\frac{[H^+][HCO_3^-]}{[H_2CO_3]} = K_r \tag{5}$$

where $K_r = 1,72*10^{-4}$ at 25°C and unit activity.

Due to the difficulty associated with specifically measuring H_2CO_3, and as only a small fraction of the total CO_2 dissolving into water is hydrolysed to H_2CO_3 (a fraction which is virtually unaffected by temperature and pH) an apparent first dissociation constant can be written incorporating the sum of the concentrations of molecularly dissolved CO_2 and H_2CO_3, i.e.

$$\frac{[\text{H}^+][\text{HCO}_3^-]}{([\text{CO}_2] + [\text{H}_2\text{CO}_3])} = \text{K}_1 \qquad (6)$$

where K_1 is the apparent first dissociation constant usually referred to as "the first dissociation constant for the carbonic system'.

Let $[\text{H}_2\text{CO}_3^*]$ refer to the sum of $[\text{H}_2\text{CO}_3]$ and molecularly dissolved CO_2, $[\text{CO}_2]$, in solution, then:

$$\frac{[\text{H}^+][\text{HCO}_3^-]}{[\text{H}_2\text{CO}_3^*]} = \text{K}_1 \qquad (7)$$

where $\text{K}_1 = 4,45.10^{-7}$ at 25°C and unit activity.

It is now possible to determine the fraction of $[\text{H}_2\text{CO}_3^*]$ existing as $[\text{H}_2\text{CO}_3]$.

Dividing Eq. (7) by Eq. (5):

$$\frac{[\text{H}_2\text{CO}_3]}{[\text{H}_2\text{CO}_3^*]} = \frac{\text{K}_1}{\text{K}_r} = \frac{4,45.10^{-7}}{1,72.10^{-4}} = 0,0026 \qquad (8)$$

That is, at 25°C only 0,26 per cent of $[\text{H}_2\text{CO}_3^*]$ exists as $[\text{H}_2\text{CO}_3]$ and 99,74 per cent as $[\text{CO}_2]$.

The second dissociation constant, K_2, for the dissociation of HCO_3^-, reaction (3), is:

$$\frac{[\text{CO}_3^=][\text{H}^+]}{[\text{HCO}_3^-]} = \text{K}_2 \qquad (9)$$

where $\text{K}_2 = 4,69.10^{-11}$ at 25°C and unit activity.

The dissociation equation of water, reaction (4), is:

$$[\text{H}^+][\text{OH}^-] = \text{K}_\text{w} \qquad (10)$$

where $\text{K}_\text{w} = 10^{-14}$ at 25°C and unit activity.

The total carbonic species concentration in solution, C_T, is defined as:

$$C_T = [H_2CO_3^*] + [HCO_3^-] + [CO_3^=] \tag{11}$$

Log [Species] - pH Diagram for the Carbonic System

Once the total carbonic species concentration, C_T, of a solution is known, equations relating the concentrations of each of the carbonic species with the hydrogen ion concentration are derived for equilibrium conditions:

Solving for $[H_2CO_3^*]$ and $[CO_3^=]$ in Eqs.(7) and (9):

$$[H_2CO_3^*] = [H^+][HCO_3^-]/K_1 \tag{12}$$

$$[CO_3^=] = K_2[HCO_3^-]/[H^+] \tag{13}$$

Substituting Eqs. (12) and (13) into Eq. (11):

$$C_T = \frac{[H^+][HCO_3^-]}{K_1} + [HCO_3^-] + \frac{K_2[HCO_3^-]}{[H^+]}$$

and simplifying

$$[HCO_3^-] = \frac{C_T}{\{[H^+]/K_1 + K_2/[H^+] + 1\}} \tag{14}$$

For convenience, define

$$\frac{[H^+]}{K_1} + \frac{K_2}{[H^+]} + 1 = X \tag{15}$$

Substituting Eq. (15) into Eq. (14):

$$[HCO_3^-] = C_T/X \tag{16}$$

Substituting Eq. (16) into Eqs. (12) and (13):

$$[H_2CO_3^*] = \frac{[H^+]}{K_1} \cdot \frac{C_T}{X} \tag{17}$$

$$[CO_3^=] = \frac{K_2}{[H^+]} \cdot \frac{C_T}{X} \tag{18}$$

The inter-relationship between the carbonic and water species with pH is conveniently depicted by means of the log [species] - pH diagram. To plot the diagram the logarithms of Eqs. (16, 17 and 18) are taken:

$$\log_{10} [HCO_3^-] = \log_{10}C_T - \log_{10}X \tag{19}$$

$$\log_{10} [H_2CO_3^*] = \log_{10}C_T - \log_{10}X + \log_{10}([H^+]/K_1) \tag{20}$$

$$\log_{10} [CO_3^=] = \log_{10}C_T - \log_{10}X + \log_{10}(K_2/[H^+]) \tag{21}$$

These equations incorporate $[H^+]$; the value of $[H^+]$ is related to pH as follows:

$$pH = -\log_{10}[H^+]$$

i.e.

$$[H^+] = 10^{-pH} \tag{22}$$

To plot the $[H^+]$ species, take the logarithm of Eq. (22):

$$\log_{10}[H^+] = -pH \tag{23}$$

The $[OH^-]$ species is plotted from the following development of Eq. (10):

Solving for $[OH^-]$ gives:

$$[OH^-] = K_w/[H^+]$$

i.e.

$$\log_{10} [OH^-] = \log_{10}K_w - \log_{10}[H^+]$$

$$= \log_{10}K_w + pH$$

but $\log_{10}K_w = -14$ at 25°C and unit activity, hence

$$\log_{10} [OH^-] = -14 + pH \tag{24}$$

In Figure 1 is shown a plot of the log of the molar concentrations of each of the species with pH (i.e. Eqs. (19 to 21) and Eqs. (23) and (24)) for a total carbonic species concentration of 0,01 moles/litre.

The approximate method for plotting the log [species] - pH diagram has been set out in Chapter 2. For the carbonic acid system there are two dissociation reactions. In constructing the diagram each dissociation is treated independently as for a monoprotic weak acid or base and the appropriate species line common to both dissociations is simply linked up. For example, in the carbonic system the HCO_3^- species is common to the dissociation equations for K_1 and K_2 and is appropriately linked as in Figure 1.

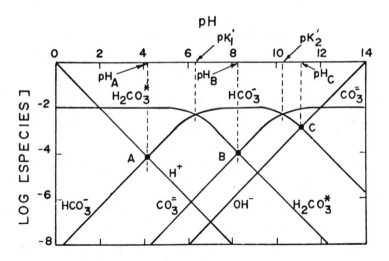

Figure 1. Distribution of carbonic species with pH; $C_T = 10^{-2}$ M

Influence of Temperature and Ionic Strength

Two factors are of importance influencing the equilibrium concentrations of the species in the carbonic system: (1) Temperature and (2)Ionic Strength.

Temperature

Dissociation constants K_1, K_2 and K_w are temperature dependent.
Effect of temperature on pK_1, Eq. (25), was determined by
Shadlovsky and McInnes (1935); on pK_2, Eq. (26) by Harned and
Scholes (1941); and on pK_w, Eq. (27) by Harned and Hamer (1933).

$$pK_1 = (17\ 052/T) + 215,21\ \log_{10}T - 0,12675\ T - 545,56 \qquad (25)$$

where T is degrees Kelvin = 272 + deg.C and the equation was
determined for the range 0°C to 38°C.

$$pK_2 = (2902,39)/T + 0,02379\ T - 6,498 \qquad (26)$$

determined in the range 0°C to 50°C. T is degrees Kelvin.

$$pK_w = 4787,3/T + 7,1321\ \log T + 0,010365\ T - 22,801 \qquad (27)$$

determined in the range 0°C to 60°C. T is degrees Kelvin.

There is a lack of data for temperatures higher than the
upper limits in Eqs. (25) to (27). In this monograph the K
values at the higher temperatures were calculated by assuming
the equations above applied. Plots of pK_1, pK_2 and pK_w for
temperatures in the range 0°C to 90°C are shown in Figure 2,
and it is observed that as temperatures increases all the pK
values decrease, i.e. the K values increase.

The log [species] – pH diagram for two waters, at 25°C and
90°C respectively, are compared in Figure 3. Each water has a
total species concentration of 0,01 moles/litre and unit activity. The diagrams are plotted using Eqs. (19) to (24) and the
appropriate K values from Eqs. (25) to (27). The respective pK
values for the two waters are listed in Table 1.

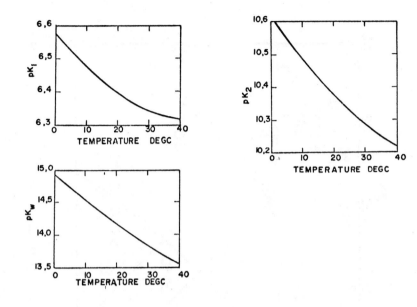

Figure 2.   Temperature dependence of pK, pK_2 and pK_w

Table 1.

Temperature	pK_1	pK_2	pK_w
25°C	6,37	10,33	14,00
90°C	6,33	10,13	12,45

From Figure 3, the most significant effect of high temperatures is the appreciable change of the pH defining neutrality (given by the intersection of the H^+ and OH^- lines), viz. pH 7,0 at 25°C to pH 6,23 at 90°C.   Also, only in the pH range 9,5 to 10,5 is there a significant change in the distribution of the carbonic species.

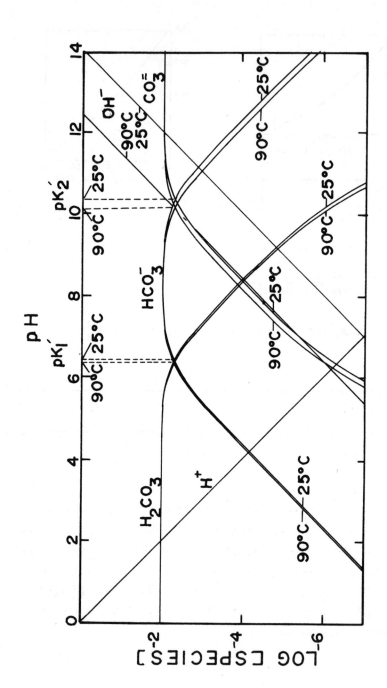

Figure 3. Effect of temperature on species distribution with pH

Ionic Strength

Increase in ionic strength reduces the activity of the species. Consequently the equilibria equations developed so far, Eqs. (7, 9 and 10), are only correct if written in terms of active concentrations, i.e.

Active concentration = activity coefficient x molar concentration.

It is convenient to retain molar concentrations in the equilibria equations. This can be achieved by suitably modifying the dissociation constants by transferring the activity coefficients to the right-hand side of the equilibria equations. The value of this dissociation constant is now the product of K and the activity factors.

A practical difficulty is determining the activity coefficients. Activity coefficients are readily determined in terms of the ionic strength by means of the Davies equation (Eq.96 Chap. 2). However, the ionic strength determination presents a problem. In Chapter 2 a procedure was set out for determining the ionic strength if the ionic constituents of the water are known. To determine the ionic strength by this procedure implies extensive chemical analysis which is not practical in the field. Fortunately, the activity coefficients are not very sensitive to ionic strength so that if only an approximate estimate of ionic strength is available, the activity factors can be determined with a degree of accuracy sufficient for most water treatment problems. Langelier (1936) established experimentally that in natural waters the ionic strength, μ, is closely estimated from the total inorganic dissolved solids concentration, S_D:

$$\mu = 2,5 \cdot 10^{-5} S_D \tag{28}$$

where μ = ionic strength in moles/litre. S_D = total inorganic

dissolved solids in mg/ℓ. This relationship is valid only for
values of S_D up to 1000 mg/ℓ.

Once the ionic strength is known the activity factors for
mono- and divalent ions of the carbonic system can be calculated
either from the Davies equation or other similar formulae (see
Chapter 2). Activity factors for mono- and divalent ions, f_m and
f_d respectively, versus total inorganic dissolved solids, S_D, are
plotted in Figure 4. The plots are based on Langelier's equation, Eq. (28), and the Davies equation, (Eq.(96) Chapter 2).

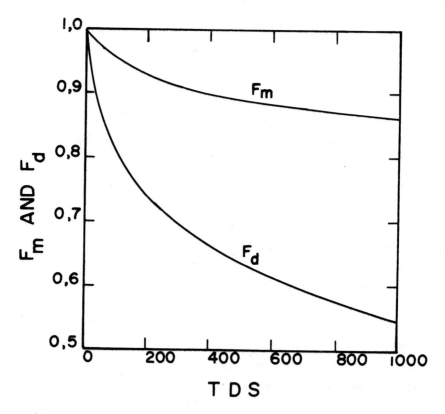

Figure 4. Variation of mono- and divalent activity coefficients
with TDS using equations of Davies and Langelier

Equations (7, 9 and 10) can now be modified to incorporate the activity factors:

$$\frac{f_m[H^+] \cdot f_m[HCO_3^-]}{[H_2CO_3^*]} = K_1 \tag{29}$$

$$\frac{f_m[H^+] \cdot f_d[CO_3^=]}{f_m[HCO_3^-]} = K_2 \tag{30}$$

$$f_m[H^+] \cdot f_m[OH^-] = K_w \tag{31}$$

where K refers to the value at a particular temperature. Note that $[H_2CO_3^*]$ is not influenced by the ionic strength as H_2CO_3 and CO_2 in solution are un-ionized.

The terms $f_m[HCO_3^-]$, $f_d[CO_3^=]$, etc. represent the active concentrations; to retain the concentrations in the molar form, transfer the f values to the right hand sides of the equations giving:

$$\frac{[H^+][HCO_3^-]}{[H_2CO_3^*]} = \frac{K_1}{f_m^2} = K_1' \tag{32}$$

$$\frac{[H^+][CO_3^=]}{[HCO_3^-]} = \frac{K_2}{f_d} = K_2' \tag{33}$$

$$[H^+][OH^-] = \frac{K_w}{f_m^2} = K_w' \tag{34}$$

However, Eq. (34) is not useful in the form K_w'. This will be apparent later.

The equations for plotting the log [species] - pH diagram for $H_2CO_3^*$, HCO_3^- and $CO_3^=$ corresponding to Eqs. (16, 17 and 18) are now:

$$\log_{10} [H_2CO_3^*] = \log_{10}C_T - \log_{10}X' + \log_{10}([H^+]/K_1') \qquad (35)$$

$$\log_{10} [HCO_3^-] = \log_{10}C_T - \log_{10}X' \qquad (36)$$

$$\log_{10} [CO_3^=] = \log_{10}C_T - \log_{10}X' + \log_{10}(K_2'/[H^+]) \qquad (37)$$

where $X' = \dfrac{[H^+]}{K_1'} + \dfrac{K_2'}{[H^+]} + 1$

Although Eqs. (35 to 37) have a form identical to Eqs. (16 to 18), the identify is deceptive when it is intended to use them to plot the log [species] - pH diagram. The reason for this is that pH is defined (and measured) as the negative log of the *active hydrogen ion concentration*, i.e.

$$pH = -\log (f_m[H^+]) \qquad (38)$$

Hence, in Eqs. (32 to 34), for any selected pH the $[H^+]$ term must be developed from Eq. (30) transformed as follows:

$$[H^+] = 10^{-pH}/f_m \qquad (39)$$

The $[H^+]$ species lines are plotted from Eq. (39), i.e.

$$\log [H^+] = -pH - \log f_m \qquad (40)$$

The $[OH^-]$ lines are plotted from Eq. (31):

$$f_m [OH^-] \cdot f_m[H^+] = K_w$$

$$\log f_m + \log [OH^-] - pH = -pK_w \qquad (41)$$

$$\log [OH^-] = pH - pK_w - \log f_m$$

It is now apparent that K_w' in Eq. (34) is not a useful form. Due to the definition of pH it is preferable to retain K_w and the form in Eq. (31).

It is of interest to note that for any ionic strength the pH of the point of intersection of $[H^+]$ and $[OH^-]$ which defines neutrality ($[OH^-] = [H^+]$) does not change. This is established as follows:

Taking the logarithm of Eq. (31):

$$\log(f_m[H^+]) + \log(f_m[OH^-]) = \log K_w$$

for neutrality $[H^+] = [OH^-]$

i.e.

$$2 \log (f_m[H^+]) = \log K_w$$

but $pH = -\log (f_m[H^+])$

i.e.

$$2 \ pH = pK_w = 14,0 \ (\text{at } 25°C)$$

i.e.

$$pH = 7,0$$

Equations (35 to 37, 40 and 41) constitute general equations for the log [species] – pH diagram of a water for any ionic strength and temperature likely to be found in treatment of surface waters. In applying temperature and ionic strength adjustments to the equilibria equations two steps are involved: first, calculate the K value for the relevant temperature for

unit activity, and second, incorporate the effect of ionic
strength as set out in Eqs. (32 to 34).

The effect of ionic strength on the species distribution of
the carbonic system is illustrated in Figure 5 by comparing the
log [species] - pH plot for two waters both at 25°C and C_T of
0,01 moles/litre but with ionic strengths 0,001 and 0,1. These
ionic strengths correspond to total dissolved solids concentrations of 40 mg/ℓ and 4 000 mg/ℓ respectively, calculated from
Langelier's formula, Eq. (28).

PRACTICAL METHODS FOR DETERMINING THE CARBONIC SYSTEM SPECIES IN WATER

In the section above it was shown that if the total carbonic
species concentration, temperature and ionic strength are
known, the concentrations of all the species constituting the
carbonic system can be calculated for any pH. However, in
practice, the operator is presented with a water of which
nothing quantitative is known. It is now necessary to determine the relevant properties of the water. For this purpose a
series of tests are performed and from these results the relevant information is obtained.

Accurate determination of the relevant properties of the
water is all important. Without this information no procedures
of treatment or control of quality can be efficiently applied.
It is the intention in this section to discuss in depth the
problems associated with the practical determination of the
constitution of the water insofar as it applied to the carbonic
system.

To determine the properties of a water the following tests
will usually be done:

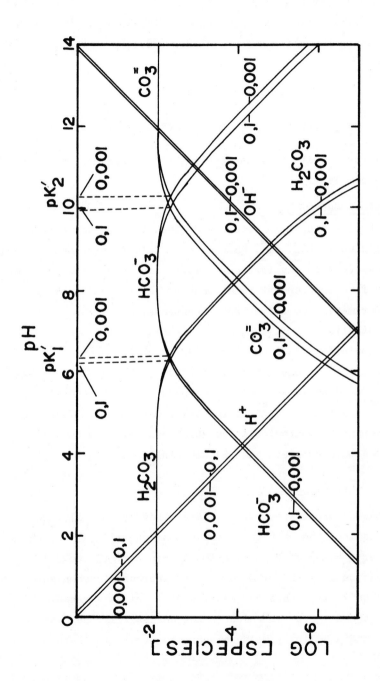

Figure 5. Effect of ionic strength on distribution of species with pH

1. Ionic strength determination, usually by means of measuring
   the total dissolved solids and finding the ionic strength
   by using Langelier's formula (Eq. (28)).

2. Temperature.

3. pH.

4. Some test to determine the total carbonic species

From the results of these tests the carbonic system can be
defined completely.

Theoretical Background

The carbonic species in water is defined by five basic parameters: $H_2CO_3^*$, HCO_3^-, $CO_3^=$, OH^- and H^+. To obtain values for
the parameters in any particular water it is necessary to establish five independent relationships (equations) linking the
basic parameters to solve these equations.

To establish the five equations, three are available from
the dissociation equations (Eqs. (32 to 34)) and one is available from the pH measurement (Eq. (38)). The remaining equation is found by establishing some measurable characteristic
which is defined in terms of the basic parameters. There are
two independent characteristics which can be measured: (1)
total carbonic species, and (2) alkalinity or acidity.

1. Total Carbonic Species Measurement

The total carbonic species, C_T, is defined by Eq. (11). Once
C_T is obtained the carbonic species concentration at the pH of
the water (or any other pH) is found from Eqs. (32 to 34).

C_T can be measured by means of the inorganic carbon analyser.
The carbon analyser has the merit that it gives a measure only
of the carbonic species which is not necessarily the case when
acidity and alkalinity measurements are used. It is therefore

very useful, indeed almost mandatory in isolating the carbonic system in waters such as digestor liquids which in addition to the carbonic system also contain the fatty acids, ammonia and phosphates. These substances can cause considerable error when utilizing the second method (below) - alkalinity titrations - to determine the carbonic system species.

The determination of the total carbonic species by means of the carbon analyser has the following disadvantages.

The analyser is as yet an expensive and temperamental instrument so that not all laboratories can afford to purchase and maintain one; the C_T determination is only really possible in a laboratory so that it has little application in field work.

2. Alkalinity and Acidity Measurement

In general alkalinity implies the existence of a net strong base which, together with an equivalent carbonic species solution (either $H_2CO_3^*$ or HCO_3^- or $CO_3^=$), establishes an observed pH. To measure alkalinity by titration, a standard strong acid is added to a water sample until the pH just equals one of the three equivalence points - the mass of strong acid added gives a measure of the alkalinity to the particular equivalence point.

Acidity implies the existence of a strong acid which, together with an equivalent carbonic species solution, establishes a pH. Acidity is therefore measured by titrating a water sample with a standard strong base to a particular equivalence point.

ALKALINITY AND ACIDITY

In Chapter 2 it was shown that diprotic acids have three equivalence points. For each of these equivalence points there is an associated alkalinity or acidity depending on whether the pH is above or below this equivalence point. Hence for the

carbonic system three forms of alkalinity and three forms of
acidity are identified - one for each of the three equivalence
points.

In Figure 1 the pH values of the three equivalence points
for the carbonic system are indicated on the log [species] - pH
diagram: pH at A corresponds to the $[H_2CO_3^*]$ equivalence point,
pH at B to the $[HCO_3^-]$ equivalence point and pH at C to the
$[CO_3^=]$ equivalence point.

Prior to the time when pH meters came into general use the
equivalence points A and B were determined approximately by
using colour pH indicators, methyl orange and phenolpthalein
respectively. Hence the custom developed to call these equivalence points the methyl orange and phenolpthalein end points;
alkalinities determined by titrating to these end points were
correspondingly called the methyl orange and phenolpthalein
alkalinities. The methyl alkalinity was also called the total
alkalinity.

In this report the following convention will apply: for alkalimetric titrations, titrations to the carbonic acid equivalence point will be termed *Alkalinity* or *Total Alkalinity*, the
two descriptions being used interchangeably. Titration to the
bicarbonate equivalence point will be designated *Phenolpthalein
Alkalinity* and titration to the carbonate equivalence point the
Caustic Alkalinity. These designations differ from those recom-
mended in Chapter 2 for a general weak acid-base system, but as
the objective of this monograph is a practical one the existing
convention for the alkalinities are retained.

The acidity measurements are defined correspondingly as
follows: titration to the carbonic acid equivalence point is
designated the *Mineral Acidity*, titration to the bicarbonate
equivalence point the CO_2-*Acidity* and titration to the carbonate
equivalence point the *Total Acidity*.

Equations for each of the three forms of alkalinity and acidity for a diprotic weak acid system were developed in Chapter 2 and are given by Eqs. (54 to 59), Chapter 2. These equations can be rewritten for the carbonic system in terms of the carbonic species as follows.

$H_2CO_3^*$ Equivalence Point

The alkalimetric titration to this equivalence point, pH_A in Figure 1, gives the Total Alkalinity (or Alkalinity) where

$$\text{Alkalinity} = 2[CO_3^=] + [HCO_3^-] + [OH^-] - [H^+] \tag{42}$$

The acidimetric titration to pH_A in Figure 1 gives Mineral Acidity where

$$\text{Mineral Acidity} = -2[CO_3^=] - [HCO_3^-] - [OH^-] + [H^+] \tag{43}$$

Mineral Acidity is also termed negative Alkalinity.

HCO_3^- (Phenolpthalein) Equivalence Point

The alkalinity titration to this equivalence point, pH_B in Figure 1, gives the Phenolpthalein Alkalinity where

$$\text{Phenolpthalein Alkalinity} = [CO_3^=] + [OH^-] - [H^+] - [H_2CO_3^*] \tag{44}$$

The acidimetric titration to the HCO_3^- equivalence point gives CO_2 Acidity where

$$CO_2 \text{ Acidity} = -[CO_3^=] - [OH^-] + [H^+] + [H_2CO_3^*] \tag{45}$$

$CO_3^=$ Equivalence Point

The alkalimetric titration to the $CO_3^=$ equivalence point, pH_C in Figure 1, gives Caustic Alkalinity where

$$\text{Caustic Alkalinity} = [OH^-] - [H^+] - [HCO_3^-] - 2[H_2CO_3^*] \tag{46}$$

The acidimetric titration to the $CO_3^=$ equivalence point gives Total Acidity (or Acidity) where

$$\text{Acidity} = -[OH^-] + [H^+] + [HCO_3^-] + 2[H_2CO_3^*] \tag{47}$$

Caustic Alkalinity is also termed negative Acidity.

The parameters Alkalinity, Acidity and pH form the bases of conditioning diagrams developed in Chapter 5. In these diagrams the terms Caustic Alkalinity and Mineral Acidity will be referred to as negative Acidity and negative Alkalinity respectively. By using this terminology only the terms Acidity and Alkalinity need be used in the formulation of the basic equations and plotting of the diagrams (see Figure 6).

Referring to Figure 1, if the initial pH is higher than C, all the alkalinities can be measured; if between B and C, only Total and Phenolpthalein Alkalinities; between A and B, only Total Alkalinity. Correspondingly, if the pH is less than A, all the acidities can be measured; between A and B, only CO_2 Acidity and Total Acidity; between C and B, only Total Acidity. It would therefore appear that Total Alkalinity and Total Acidity are the most practical parameters to measure as they are not restricted by the initial pH to the same degree as the other alkalinities and acidities. In practical water analysis Total Alkalinity is superior to Total Acidity as the carbonic acid equivalence point can be identified relatively easily, whereas the carbonate end point is usually indecisive and, for other reasons (discussed later), is liable to result in large errors in the Total Acidity evaluation.

Species Concentrations in Terms of Total Alkalinity

Once Total Alkalinity and pH of a water have been measured, the $[H_2CO_3^*]$, $[HCO_3^-]$ and $[CO_3^=]$ concentrations can be calculated as follows:

Solve for $[CO_3^=]$ from Eq. (33)

$$[CO_3^=] = K_2' \, [HCO_3^-]/[H^+] \tag{48}$$

Solve for $[H_2CO_3^*]$ from Eq. (32)

$$[H_2CO_3^*] = [H^+][HCO_3^-]/K_1' \tag{49}$$

Substituting for $[CO_3^=]$ from Eq. (48) into the equation for Alkalinity, Eq. (42) and solving for $[HCO_3^-]$:

$$[HCO_3^-] = \frac{Alkalinity - [OH^-] + [H^+]}{(1 + 2K_2'/[H^+])} \tag{50}$$

Substituting for $[HCO_3^-]$ from Eq. (50) into Eqs. (48 and 49) and simplifying:

$$[CO_3^=] = \frac{Alkalinity - [OH^-] + [H^+]}{([H^+]/K_2' + 2)} \tag{51}$$

$$[H_2CO_3^*] = \frac{Alkalinity - [OH^-] + [H^+]}{(K_1'/[H^+] + 2K_2'.K_1'/[H^+]^2)} \tag{52}$$

Total carbonic species concentration, C_T, can be expressed in terms of the parameters Alkalinity and $[H^+]$ as follows:

Substitute Eqs. (50, 51 and 52) into Eq. (11) and simplify:

$$C_T = (1 + K_2'/[H^+] + [H^+]/K_1')(\frac{Alkalinity - K_w'/[H^+] + [H^+]}{1 + 2K_2'/[H^+]}) \tag{53}$$

Relationship between Total Alkalinity, Total Acidity and Total
Carbonic Species Concentration

Examination of Eqs. (42 and 47) for Total Alkalinity and Total Acidity respectively, shows that their sum exactly equals twice the total carbonic species concentration, i.e.

$$Total\ Alk + Total\ Acidity = 2[CO_3^=] + 2[HCO_3^-] + 2[H_2CO_3^*] \tag{54}$$

$$= 2C_T$$

Thus, solving for C_T in Eq. (53) using measured values of Total Alkalinity and pH, Total Acidity can now be calculated using Eq. (54), i.e.

$$Total\ Acidity = 2C_T - Total\ Alk$$

Subtracting Eq. (47) from Eq. (42) gives a relationship linking
Alkalinity, Acidity and Phenolpthalein Alkalinity (or negative
CO_2 Acidity),

$$\text{Alkalinity} - \text{Acidity} = 2[CO_3^=] + 2[OH^-] - 2[H_2CO_3^*] - 2[H^+]$$

(55)

$$= 2 \text{ Phenolpthalein Alkalinity}$$

The inter-relationships between C_T and the various forms of
alkalinity and acidity for the carbonic system are illustrated
in Figure 6.

Total Alkalinity and pH are the two parameters most commonly
measured in order to define the carbonic system. For this
reason the equations for $[CO_3^=]$, $[HCO_3^-]$ and $[H_2CO_3^*]$, Eqs. (50 to
52), were developed in terms of these two measured values. However, the carbonic system is completely defined by measuring
any two parameters in the system and species concentrations can
be calculated using equilibria equations, Eqs. (32, 33 and 34).

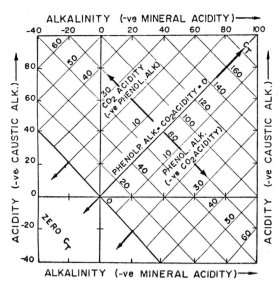

Figure 6.   Inter-relation between C_T and the various forms of
alkalinity and acidity

MEASUREMENT OF ALKALINITY

Factors Affecting Equivalence Points

The equivalence points of $H_2CO_3^*$, HCO_3^- and $CO_3^=$ are each affected
by three factors: (1) temperature; (2) ionic strength, and (3)
total carbonic species concentration.

Temperature:

Referring to Figure 3, the species lines are compared for two
waters with the same ionic strength and total carbonic species
concentration but two temperatures 25°C and 90°C respectively.
It is evident that the $H_2CO_3^*$ equivalence point is virtually unaffected by temperature, but the HCO_3^- and $CO_3^=$ equivalence points
are appreciably affected.

Ionic Strength:

Referring to Figure 5, the species lines are compared for two
waters at the same temperature, having the same total carbonic
species concentration, but having ionic strength of 0,001 and
0,1 respectively. Again the effect on the $H_2CO_3^*$ equivalence
point is negligible, but the effect on the HCO_3^- and $CO_3^=$ equivalence points is appreciable.

Total Carbonic Species Concentration:

Referring to Figure 7, two waters are compared both having the
same ionic strength and at the same temperature, but with different total carbonic species concentrations. The $H_2CO_3^*$ and
$CO_3^=$ equivalence points are strongly affected, but the HCO_3^-
point is unaffected.

Comparing the three equivalence points, there is a definite
practical advantage in titrating to an equivalence point which
is not affected by temperature and ionic strength for this implies that no correction need to be made for these two effects
in any field titration. In this respect the $H_2CO_3^*$ equivalence
point is superior to both the $CO_3^=$ and HCO_3^- points. However, an

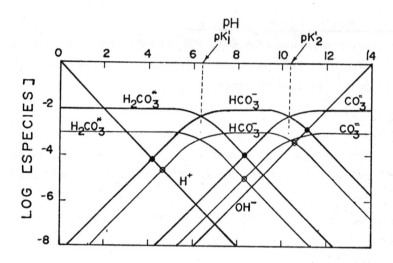

Figure 7. Effect of C_T on the distribution of species with pH

over-riding consideration, again from a practical point, is -
can the end point be identified in the field? To clarify this
aspect it is necessary to consider the buffering capacity of
the carbonic system and a comparison of the buffering capacity
and titration curves.

Practical Estimation of the pH of the Equivalence Points

Relationship between Titration Inflection Point, Buffer Capacity and Equivalence Point:

It was shown in Chapter 2 that the buffer capacity equation for
a diprotic acid can be approximated to the buffer equation for
two monoprotic acids with equal total species concentrations
provided the ratio of successive ionization constants (i.e. K_1/K_2) is greater than 100. This provision is valid for the

carbonic system. Hence the buffer equation for the system approximates to:

$$\beta_T = \beta_{H_2O} + \beta_{H_2CO_3^*} + \beta_{HCO_3^-} \qquad (56)$$

where β_T is the total buffer capacity and β_{H_2O}, $\beta_{H_2CO_3^*}$ and $\beta_{HCO_3^-}$ are the components of β_T due to the dissociation of water, $H_2CO_3^*$ to HCO_3^- and HCO_3^- to $CO_3^=$ respectively:

$$\beta_{H_2O} = 2,303([H^+] + [OH^-]) \qquad (57)$$

$$\beta_{H_2CO_3^*} = \frac{2,303 \cdot C_T \cdot K_1' \cdot [H]}{(K_1' + [H^+])^2} \qquad (58)$$

$$\beta_{HCO_3^-} = \frac{2,303 \cdot C_T \cdot K_2' \cdot [H^+]}{(K_2' + [H^+])^2} \qquad (59)$$

Another useful expression for $\beta_{H_2CO_3^*}$ is obtained by substituting Eqs. (11 and 32) and simplifying:

$$\beta_{H_2CO_3^*} = \frac{2,303 \cdot [H_2CO_3^*][HCO_3^-]}{[H_2CO_3^*] + [HCO_3^-]} \qquad (60)$$

Similarly, for HCO_3^-, by substituting Eqs. (11 and 33) into Eq. (59) and simplifying:

$$\beta_{HCO_3^-} = \frac{2,303 \cdot [HCO_3^-][CO_3^=]}{[HCO_3^-] + [CO_3^=]} \qquad (61)$$

The total buffer capacity, β_T, and its components β_{H_2O}, $\beta_{H_2CO_3^*}$ and $\beta_{HCO_3^-}$ are plotted against pH for C_T 10^{-3} moles/ℓ in Figure 8. Integrating the buffer capacity curve from say pH 1 to pH 12, the theoretical titration-pH curve is obtained and

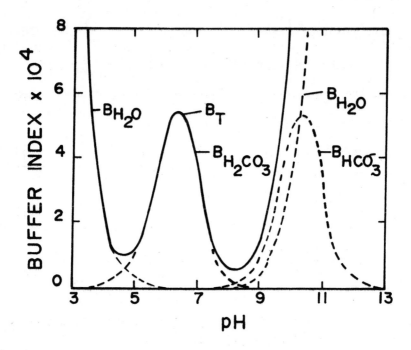

Figure 8. Variation of buffer capacity, β, with pH. $C_T = 10^{-3}M$

the slope at any point on this curve is equal to the buffer
capacity (Figure 9).

The buffer capacity curve shows marked minima at pH values
of about 4,5 and 8,4 and a minor minimum at pH about 10,4. At
these minima the pH-titration curve should show points of inflection of curvature and relatively low slope values. Low
slopes and points of inflection are discernible at pH 4,5 and
8,4 but not at pH 10,4. The reason for this is that the buffer
capacity is large at pH 10,4 due to both the β_{H_2O} and $\beta_{HCO_3^-}$
components whereas at 4,5 and 8,4 the buffering capacities are
small. The distinctiveness of the points of minimum buffer

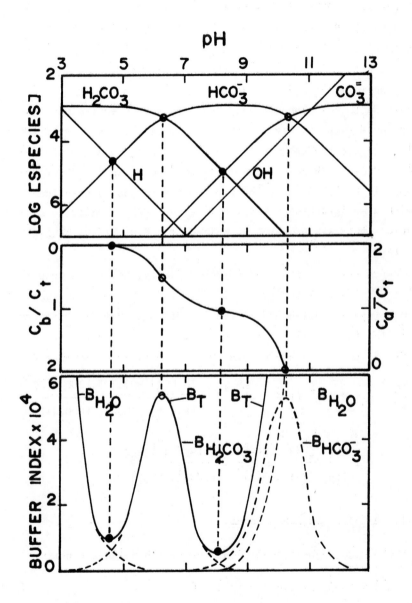

Figure 9.   Interdependence between log [species] - pH diagram
titration curve and buffer capacity diagram

capacity depends in large measure on the magnitude of the buffer
capacity at that point compared with the buffer capacity at
nearby pH values. Consequently, points of minimum buffer capacity at pH 4,5 and 8,4 are clearly defined in the titration
curve but unsatisfactorily or not at all, at pH 10,4.

The inflection points at pH about 4,5 and 8,4 appear to correspond to the carbonic acid and bicarbonate equivalence points
(A and B respectively) in the log [species] - pH diagram (Figure
1). As the titration curve can be established experimentally,
if one could show that the inflection point in fact corresponds
exactly to the pH's of the equivalence points, a practical
method of determining these equivalence points would be available. This can be shown as follows for each of the equivalence
points:

From the definition of buffer capacity, Eq. (63), Chapter 2,
the pH of inflection points in the titration curve corresponds
to points of maxima and minima in buffer capacity. Differentiating the equation for buffer capacity, Eq. (56), with respect to pH and equating to zero for a maximum or minimum gives:

$$\frac{\delta\beta}{\delta pH} = 0 = 2,303^2.([OH^-] - [H^+] - \frac{[H_2CO_3^*][HCO_3^-]}{[H_2CO_3^*]+[HCO_3^-]} - \frac{[HCO_3^-][CO_3^=]}{[HCO_3^-]+[CO_3^=]}$$

$$+ \frac{2.[H_2CO_3^*]^2.[HCO_3^-]}{([H_2CO_3^*]+[HCO_3^-])^2} + \frac{2.[HCO_3^-]^2[CO_3^=]}{([HCO_3^-]+[CO_3^=])^2}) \qquad (62)$$

(i) H_2CO_3^* Equivalence Point:

At the $H_2CO_3^*$ equivalence point the proton balance equation is:

$$2[CO_3^=] + [HCO_3^-] + [OH^-] - [H^+] = 0 \qquad (63)$$

At the point of minimum buffer capacity in this pH region:
$[H_2CO_3^*] \gg [HCO_3^-] \gg [CO_3^=]$ and $\delta\beta/\delta pH$ equals zero. Equation
(62) approximates to $[CO_3^=] + [HCO_3^-] + [OH] - [H^+] = 0$, which

approximates to the proton balance equation for the $H_2CO_3^*$ equivalence point, Eq. (63). Thus the $H_2CO_3^*$ equivalence point corresponds to an inflection point in the titration curve.

(ii) HCO_3^- Equivalence Point:

For an equivalent HCO_3^- solution, the proton balance equation is

$$[CO_3^=] + [OH^-] - [H^+] - [H_2CO_3^*] = 0 \qquad (64)$$

At the point of minimum buffer capacity in this pH region: $[HCO_3^-] \gg [H_2CO_3^*]$ or $[CO_3^=]$, and $\delta\beta/\delta pH$ equals zero. Equation (62) approximates to $[OH] - [H^+] - [H_2CO_3^*] + [CO_3^=] = 0$, which is the same equation as Eq. (64) showing that the HCO_3^- equivalence point corresponds to a minimum in the buffer capacity and an inflection point in the titration curve.

(iii) $CO_3^=$ Equivalence Point:

The proton balance equation for an equivalent $CO_3^=$ solution is:

$$2[H_2CO_3^*] + [HCO_3^-] + [H^+] - [OH^-] = 0 \qquad (65)$$

If a point of minimum buffer occurs in this region then $\delta\beta/\delta pH$ equals zero, and if at this point $[CO_3^=] \gg [HCO_3^-]$, then Eq. (62) approximates to the proton balance equation for $CO_3^=$, Eq. (65), and an inflection point occurs in the titration curve at this pH.

However, as C_T decreases, the $CO_3^=$ equivalence point decreases and the $[CO_3^=]$ and $[HCO_3^-]$ concentrations approach the same order of magnitude at the equivalence point. When C_T equals about $10^{-3,4}$ moles/ℓ, the equivalence point occurs at pH equal to pK_2. At this point $[HCO_3^-]$ equals $[CO_3^=]$ and Eq. (62) approximates to

$$\frac{\delta\beta}{\delta pH} = ([OH^-] - [H^+]) \cdot 2,303^2$$

Thus, $\delta\beta/\delta pH$ no longer equals zero at the equivalence point and no inflection point occurs in the titration curve.

In general for C_T less than about 10^{-2} moles/ℓ, no clear

inflection point will be observed at the $CO_3^=$ equivalence point
and no titration to this endpoint is practical. Where C_T is
greater than 10^{-2} moles/ℓ the inflection point is discernible,
but the correspondingly high buffer capacity causes unacceptably
high titration errors at this endpoint. In most terrestrial
waters C_T is less than 10^{-2} moles/ℓ so that an inflection point
is not observed.

Titration Errors

The error in alkalinity or acidity measurement resulting from
titration to an incorrect endpoint was set out in Chapter 2.
Equations for errors in titration to each of the three equivalence points for the carbonic system are presented below:

(i) $\underline{H_2CO_3^* \ endpoint}$

Titration error $\approx 4,606.[H^+]_e.\Delta pH$

Where $[H^+_e]$ is the hydrogen ion concentration at the
equivalence point.

ΔpH is the pH error in the titration endpoint

The titration error (in moles/ℓ) at this endpoint in
terms of C_T and $[H^+]_e$ is

$$Error = 2,303.\left([H^+]_e + \frac{C_T K_1}{[H^+]_e}\right).\Delta pH$$

(ii) $\underline{HCO_3^- \ (Phenolpthalein) \ endpoint}$

$$Titration \ error = 2,303.\left([H^+]_e + [OH^-]_e + \frac{C_T[H^+]_e}{K_1} \right.$$
$$\left. + \frac{C_T.K_2}{[H^+]_e}\right).\Delta pH \ (moles/\ell)$$

Where $[OH^-]_e$ is the hydroxyl ion concentration at the
equivalence point.

(iii) $\underline{CO_3^=}$ endpoint

Titration error = $4{,}606 \cdot [OH^-]_e \cdot \Delta pH$ (moles/ℓ)

and in terms of C_T and $[H^+]_e$

$$\text{Error} = 2{,}303 \cdot \left(\frac{K_w}{[H^+]_e} + \frac{C_T \cdot K_2 \cdot [H^+]_e}{(K_2 + [H^+]_e)^2} \right) \cdot \Delta pH \text{ (moles/ℓ)}$$

Selection of Titration Endpoint

The theory of the behaviour at each equivalence point has been set out in detail. From a practical point of view the summary below lists the advantages and disadvantages of measuring each of the various forms of alkalinity and acidity by titrating a water sample to the respective equivalence point.

$\underline{CO_3^=}$ Equivalence Point:

The pH at which this point occurs is affected by temperature, ionic strength and C_T; the inflection point is not easily (if at all) identifiable in the titration curve; large titration errors result from the high buffer capacity at the endpoint of the titration. A further disadvantage of titrating a sample to this endpoint is that as the acidity titration progresses the $CO_3^=$ and OH^- concentrations increase and may reach a level where relatively insoluble calcium and magnesium salts precipitate, causing a titration error.

$\underline{HCO_3^-}$ Equivalence Point:

The pH of this point is affected relatively strongly by both temperature and ionic strength, but is independent of C_T. A major disadvantage of measuring alkalinity or acidity to this point is that most terrestrial waters have a pH close to this equivalence point, consequently low (or sometimes zero) titrations result. A low titration results in a large relative error

even though the buffer capacity (and hence absolute error) is
low. A zero titration gives no additional information regarding
the system over and above the measured pH. Because this point
is independent of C_T a measured pH of 8,4 and a zero Phenolpthalein titration give only *one* item of information.

$H_2CO_3^*$ Equivalence Point:

The pH of this point is affected minimally by temperature and
ionic strength. However it has the disadvantage that it is affected by C_T. This means that for different samples the endpoint will vary depending on C_T and C_T is initially unknown. A
further disadvantage of an alkalinity titration to this endpoint
is that $[H_2CO_3^*]$ (and hence the concentration of molecularly dissolved CO_2) increases as the titration progresses. This increases the possibility of CO_2 being lost from the sample to
the atmosphere. Such CO_2 loss would change C_T and hence the pH
of the endpoint. A titration to a predetermined endpoint would
thus give incorrect results; a titration to the corresponding
inflection point in the titration curve would *not* cause an error
in Alkalinity measurement as CO_2 loss does not directly affect
Alkalinity (see Chapter 5). In general, provided that the titration is carried out with smooth stirring (using say a magnetic stirrer) CO_2 losses are negligible. The low value for the
minimum buffer capacity associated with this endpoint results
in a clearly defined inflection point in the titration curve
and a corresponding low titration error.

Comparing the practicality of titrating to these three equivalence points, there is a definite advantage in measuring the
parameter Total Alkalinity by titrating to the $H_2CO_3^*$ endpoint.
This means however, that the endpoint must be defined for each
titration.

Determination of the Carbonic Acid Equivalence Point by Direct Titration between Preselected pH Values

A practical method of estimating the pH of the $H_2CO_3^*$ equivalence in the field is as follows. Adjust the pH of the water sample to 6,0 using a standard strong acid. Measure the alkalinity required to titrate the water sample from pH 6,0 down to 4,5. The $H_2CO_3^*$ equivalence point for the water is now obtained directly by referring the measured alkalinity from pH 6,0 to 4,5 value to the plot given in Figure 10.

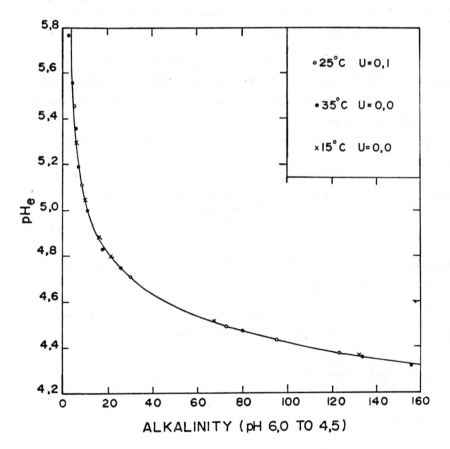

Figure 10. Variation of $H_2CO_3^*$ equivalence point, pH_e, with alkalinity between pH 6,0 and 4,5

The development of this method of endpoint estimation is based on the following theoretical considerations:

(i)   the pH of the $H_2CO_3^*$ equivalence point is dependent principally on the total carbonic species concentration and is virtually independent of ionic strength and temperature.

(ii) by measuring the alkalinity between any two fixed pH values, C_T can be closely calculated.

(iii) for any C_T value the $H_2CO_3^*$ equivalence point pH can be calculated from basic theory.

(i) The observation that the $H_2CO_3^*$ equivalence point is virtually unaffected by both temperature and ionic strength (Figures 3 and 5) and depends only on C_T (Figure 7) is established theoretically as follows:

At the $H_2CO_3^*$ equivalence point $\delta\beta/\delta pH$ is zero, hence from Eq. (62):

$$\frac{\delta\beta}{\delta pH} = 0 \approx 2{,}303\ ([OH]_e - [H^+]_e - \frac{C_T K_1'[H^+]_e}{(K_1'+[H^+]_e)^2} + \frac{2C_T K_1'[H^+]_e^2}{(K_1'+[H^+]_e)^3}\ )$$

$$(66)$$

where subscript 'e' refers to species concentrations at the $H_2CO_3^*$ equivalence point.

Also at this equivalence point $[H^+]_e \gg K_1$ for C_T greater than about 10^{-4} M and Eq. (66) approximates to:

$$[H^+]_e + \frac{C_T \cdot K_1'}{[H^+]_e} - \frac{2C_T \cdot K_1'}{[H^+]_e} = 0$$

Solving for $[H^+]_e$

$$[H^+]_e = \sqrt{C_T K_1'}$$

Solving for pH_e

$$f_m \cdot [H^+]_e = f_m \cdot \sqrt{C_T} \cdot K_1'$$

$$= \sqrt{C_T} K_1$$

$$pH_e = 1/2 \cdot pC_T + 1/2 \; pK_1 \tag{67}$$

Equation (67) is in a form in which the relative effects can be estimated of ionic strength, temperature and total carbonic species concentration on the $H_2CO_3^*$ equivalence point.

In Eq. (67) ionic strength has no effect on the $H_2CO_3^*$ equivalence point pH, pH_e, as ionic strength effects cancel out to leave pK_1 in the thermodynamic form. The effect of temperature on pH_e is negligible. This is appreciated by noting that pK_1 (the term which incorporates the temperature effect on pH_e) only varies from 6,37 at 25°C to 6,33 at 90°C (see Figure 2). The effect of pC_T on pH_e is significant. In Eq. (67) pH_e will change proportionally to change in pC_T and pC_T may have a value in the range 4,5 down to 1,5 for natural waters.

(ii) By measuring the alkalinity of a water between any two fixed pH values the total carbonic species concentration, C_T, can be calculated from basic theory. Then, using the relationship between C_T and the $H_2CO_3^*$ equivalence point, Eq. (67), one can solve for the $H_2CO_3^*$ equivalence point pH in terms of the alkalinity between the two fixed pH values.

(iii) The theoretical relationship between alkalinity (from the two fixed pH values, say pH 6,0 and 4,5) and C_T is obtained as follows: For a particular C_T value the buffer capacity-pH curve is plotted from pH 6,0 down to 4,5 using Eqs. (56 to 59). The cumulative integral under this buffer capacity-pH curve gives the theoretical titration-pH curve between pH 6,0 and 4,5 for

the particular C_T in question. The $H_2CO_3^*$ equivalence point pH
for this C_T value is then obtained from Eq. (67). Repeating
this procedure for a number of C_T values, the theoretical plot
of alkalinity from pH 6,0 to 4,5 against the $H_2CO_3^*$ equivalence
point pH is obtained (Figure 10).

The theory above provides a practical method for field and
laboratory to obtain accurate Alkalinity measurements. Steps
to be followed are (1) adjust the pH of the sample to pH 6,0;
(2) titrate down to pH 4,5 and note alkalinity measured. In
Figure 10 read endpoint pH from measured alkalinity. (3) Titrate a new sample of the water to be tested to the endpoint pH
determined in (2) to obtain Total Alkalinity. No adjustment is
necessary for ionic strength and temperature.

Precautions in Alkalinity and Acidity Titrations

The chemistry of $CaCO_3$ and CO_2 solubility in water will be discussed in Chapters 4 and 5 respectively. However, because
these two parameters can significantly affect the accuracy of
alkalinity and acidity titrations, a brief discussion is given
below on their possible effects.

Effect of CO_2 Loss or Gain on Alkalinity and Acidity Titrations

The tendency for CO_2 exchange to occur between water and air
depends on the relative CO_2 partial pressures in these two
phases. If the partial pressures are equal, no CO_2 exchange
occurs. If however, the CO_2 partial pressure in the water phase
is greater than in the air, there is a possibility that CO_2 will
be expelled from the water causing a pH increase and a C_T decrease (and vice versa for the CO_2 partial pressure in air being
greater than that in the water phase).

One should appreciate that in alkalimetric titrations the
concentration of $H_2CO_3^*$ (and hence CO_2) in the water increases
as the titration progresses and there is an increasing possibility of the CO_2 partial pressure in the water exceeding that in

the air and hence for CO_2 to be expelled from the water (and vice versa for acidity titrations).

If CO_2 is lost from a water sample during a Total Alkalinity titration the resultant decrease in C_T causes an increase in the $H_2CO_3^*$ equivalence point. The Total Alkalinity value, however, does not change through loss or gain of CO_2. This is evident from the Total Alkalinity Eq. (42) which is independent of $H_2CO_3^*$. Because the endpoint changes if CO_2 is lost from a water sample during titration, any titration to a predetermined endpoint will result in an error in the Alkalinity measurement.

In acidimetric titrations there is an increasing possibility of the water sample absorbing CO_2 from the air as titration progresses. If the Total Acidity is measured by titration to a predetermined endpoint, and if the sample absorbs CO_2 from the air, then the titration endpoint pH increases as C_T increases and also the Total Acidity value of the sample increases by a concentration equivalent to the CO_2 absorbed as is evident in Eq. (47).

Weber and Stumm (1962), state that provided smooth stirring is maintained in alkalinity titrations, loss of CO_2 from the sample is negligible. However, a good experimental technique to follow is that once the pH endpoint has been established for a particular water, subsequent titrations to this endpoint should be performed using the same procedure.

Effect of $CaCO_3$ Solubility on Alkalinity and Acidity Titration

If the product of the molar concentrations of calcium and carbonate ions in solution exceeds a value known as the solubility product constant for $CaCO_3$, K_s, solid $CaCO_3$ is precipitated from the solution until the product K_s is re-achieved.

In acidimetric titrations of natural waters the $CO_3^=$ concentration increases as the titration progresses, and the condition may be reached where the $CaCO_3$ solubility product is exceeded

and $CaCO_3$ precipitated. The loss of $CO_3^=$ ions in the precipitate
decreases C_T so that the endpoint pH changes. If a Total Acidity
titration is carried out to a predetermined endpoint and $CaCO_3$
precipitation occurs, an error will result in the Acidity value.
If the titration is carried out to the correct endpoint, then
from Eq. (47) it is evident that the Total Acidity value remains
unchanged as it is independent of $CO_3^=$.

For alkalimetric titrations, if finely dispersed particles
of $CaCO_3$ are present in the water sample, then as the titration
progresses the solid $CaCO_3$ dissolves as the pH is lowered. If
the Total Alkalinity of such a water was measured by titration
to the $H_2CO_3^*$ inflection point, in the experimental titration
curve not only would the pH of this inflection point decrease
as the solid $CaCO_3$ dissolves, but also (as is evident from Eq.
(42)) the measured Total Alkalinity value would be overestima-
ted by an amount equivalent to the concentration of solid $CaCO_3$
which dissolved during the titration.

Estimation of Species Concentrations - a Criticism of Current Method

The concentrations of each of the carbonic species can be estimated provided that two independent parameters for the systems
are measured. Previous work in this chapter indicates that in
practice it is usually possible to measure three parameters for
the carbonic system - pH, Alkalinity and Phenolpthalein Alkalinity. Current practice is to estimate individual species concentrations from Alkalinity and Phenolpthalein Alkalinity
measurement using approximations to the equations for these two
parameters. Though the method is simple it can lead to serious
errors where C_T is low (below about 10^{-3} moles/ℓ).

The basic theory for this method and sources of error are
set out as follows:

Carbonic species concentrations can be determined from Alkalinity and Phenolpthalein Aklalinity measurements by making judicious assumptions as to the significance of certain species concentrations in the equations for these two parameters, Eqs. (68 and 69). (See Standard Methods, 1963).

$$\text{Alkalinity} = 2[CO_3^=] + [HCO_3^-] + [OH^-] - [H^+] \qquad (68)$$

$$\text{Phenolpthalein Alkalinity} = [CO_3^=] + [OH^-] - [H^+] - [H_2CO_3^*] \quad (69)$$

In these two equations, depending on the values of parameters measured, four situations may arise:

Case 1: Alkalinity positive; Phenolpthalein Alkalinity negative.

Referring to Figure 1 the condition of the water lies between the $H_2CO_3^*$ and HCO_3^- equivalence points. Each of the species on the right hand side of Eqs. (68 and 69) are negligible except for HCO_3^- and $H_2CO_3^*$. These two equations thus approximate to

$$T = [HCO_3^-] \qquad \text{and}$$

$$P = -[H_2CO_3^*]$$

where T and P refer to Alkalinity and Phenolpthalein Alkalinity respectively.

Thus, the *acidity* titration to the phenolpthalein endpoint will closely give a measure of $[H_2CO_3^*]$ - often referred to in the literature as 'free CO_2'.

Case 2: Alkalinity and Phenolpthalein Alkalinity both positive and pH between the HCO_3^- and $CO_3^=$ equivalence points.

Referring to Figure 1 all the species are negligible in this pH range except $CO_3^=$ and HCO_3^-. Eqs. (68 and 69) approximate to

$$P = [CO_3^=] \qquad \text{and}$$

$$T = 2[CO_3^=] + [HCO_3^-]$$

Solving for $[HCO_3^-]$

$$[HCO_3^-] = T - 2P$$

If $[HCO_3^-]$ is calculated to have a negative value then the assumption $[OH^-]$ as negligible is invalid and species concentrations are calculated as set out below.

Case 3: Alkalinity and Phenolpthalein Alkalinity positive; pH above the $CO_3^=$ equivalence point

All the species for the system are negligible in this pH region except $[CO_3^=]$ and $[OH^-]$. Eqs. (68 and 69) thus approximate to

$$T = 2[CO_3^=] + [OH^-] \qquad (70)$$

$$P = [CO_3^=] + [OH^-] \qquad (71)$$

Subtracting Eq. (71) from Eq. (70) and solving for $[CO_3^=]$

$$[CO_3^=] = T - P$$

and substituting for $[CO_3^=]$ into Eq. (71) and solving for $[OH^-]$

$$[OH^-] = 2P - T$$

Case 4: Alkalinity equals Phenolpthalein Alkalinity

The only explanation for this observation is that C_T is zero. The only species of note is $[OH^-]$, thus

$$P = T = [OH^-]$$

The above method for estimating species concentrations gives incorrect results where C_T for the water is less than about 10^{-3} M and pH is above pH 9,0, see Figure 11. In this case $[OH^-]$, $[HCO_3^-]$ and $[CO_3^=]$ are all of the same order of magnitude and none of these three species can be neglected. Consequently the species concentrations cannot be determined by the method discussed above. As $C_T < 10^{-3}$ M is common in terrestrial waters

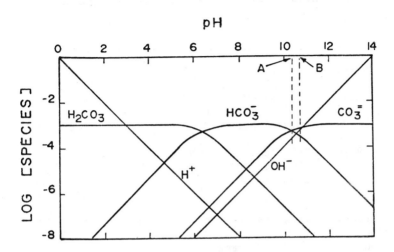

Figure 11. $[HCO_3^-]$, $[CO_3^=]$ and $[OH^-]$ concentrations have the same order of magnitude in pH region A to B

the method has restricted use.

A general semi-graphical method for estimating species concentrations is as follows: (i) calculate C_T from the measured values for Alkalinity and Phenolpthalein Alkalinity using Eqs. (54 and 55); (ii) sketch the log [species] - pH diagram for the calculated value of C_T; (iii) in the log [species] - pH diagrams read off the species concentrations at the measured pH.

SUMMARY

(1) Using equilibrium chemistry, the state of a water is defined by measuring at least two parameters.   In practice these two parameters are usually pH and Total Alkalinity.

(2) Total Alkalinity is measured by titrating a water sample to its $H_2CO_3^*$ equivalence point.   However, the pH of this

Table 2.

Parameter	$H_2CO_3^*$ endpoint	HCO_3^- endpoint	$CO_3^=$ endpoint
Ionic strength	Negligible	Appreciable	Appreciable
Temperature	Negligible	Appreciable	Appreciable
C_T	Appreciable	No effect	Appreciable
CO_2 Exchange	Tendency for CO_2 loss and endpoint pH elevated	No effect	Tendency for CO_2 absorption and endpoint pH depressed
$CaCO_3$ precipitation o solution	If solid $CaCO_3$ is present it dissolves. Endpoint pH depressed.	No effect	If precipitation occurs, endpoint pH decreases
Clarity of endpoint in titration curve	Good	Good	Bad

equivalence point varies from water to water. In Table 2 the effects on the pH of the $H_2CO_3^*$ equivalence point of various parameters such as C_T, ionic strength, temperature are summarized and compared with the effects on HCO_3^- and $CO_3^=$ equivalence points.

(3) In practice the pH of the $H_2CO_3^*$ equivalence point for a water sample can be obtained from the pH of the relevant inflection point in the titration curve. This method of endpoint evaluation is time consuming.

(4) An alternative to (3) is to titrate a water sample
     between fixed pH values (say from pH 6,0 down to
     4,5) in which event C_T can be approximately evaluated.
     The $H_2CO_3^*$ equivalence point and C_T are directly re-
     lated, hence a plot of alkalinity from pH 6,0 to 4,5
     against $H_2CO_3^*$ equivalence point is obtained (Figure
     10) and used to estimate the $H_2CO_3^*$ equivalence point.
     (In Chapter 5 yet a further method for determining
     Total Alkalinity and Total Acidity will be discussed).

References

Butler,J.N.   1964   *Ionic Equilibrium. A Mathematical
                                 Approach*, Addison Wesley, Reading,
                                 Mass.

Hamer,P., Jackson, 1961   *Industrial Water Treatment Practice*,
J, and Thurston,   Butterworth, London.
E.F.

Harned,H.S. and   1933   "The Ionization Constant of Water",
Hamer,W.J.   J.Am.Chem.Soc., 51, 2194.

Harned,H.S. and   1941   "The Ionization Constant of HCO_3^-",
Scholes,S.R.   J.Am.Chem.Soc., 63. 1706.

Shadlovsky,T.and   1935   "The First Ionization Constant of
MacInnes D. HCO_3^-", J.Am.Chem.S c., 59, 2304.

Weber,W.J. and   1963   "Mechanism of Hydrogen Ion Buffering
Stumm,W   in Natural Waters", J.A.W.W.A., 55
                            1553.

Chapter 4

EQUILIBRIUM OF A SOLID WITH ITS DISSOLVED SPECIES

Solubility of Solids in Water

When a salt dissolves in water, the mass which can dissolve per
unit volume of liquid is limited. This limiting concentration
is called the solubility of the salt in the particular liquid.
In Table 1 the solubilities in water of bicarbonates, carbonates,
chlorides, hydroxides and sulphates of the cations calcium, magnesium and sodium are listed, Nordell (1961). These values can
be considered as only relative for the conditions under which
the values were obtained are not defined. However, the table
does present a general picture of the relative solubilities of
these compounds.

From Table 1, compared with calcium and magnesium salts, all
the sodium salts are extremely soluble. In calcium salts only
calcium carbonate has an extremely low solubility. In the magnesium salts both magnesium hydroxide and magnesium carbonate
are relatively insoluble.

The control of the concentration of the cations Ca^{++} and Mg^{++}
in water treatment is of the greatest importance in boiler feed
waters, domestic water supplies and many industrial applications.
The inter-relation between the solubilities of the calcium and
magnesium compounds with the various chemical constituents in
the water is complex and quantitative solutions are only possible
using the theory of weak acid-base equilibria and solubility
products.

Solubility Product

If an excess of salt is added to a liquid, eventually a state of
dynamic equilibrium is achieved between dissolution of the solid
and precipitation of the ions to the solid state. For example,
if excess solid $Mg(OH)_2$ is added to pure water, the equilibrium
reaction is:

Table 1
Solubilities of Calcium, Magnesium and Sodium bicarbonate and
carbonate, Chloride, Hydroxide and Sulphate

Mineral	Formula	Solubility as ppm of $CaCO_3$ at 0°C
Calcium bicarbonate	$Ca(HCO_3)$	1 620
Calcium carbonate	$CaCO_3$	15
Calcium chloride	$CaCl_2$	336 000
Calcium hydroxide	$Ca(OH)_2$	2 390
Calcium sulphate	$CaSO_4$	1 290
Magnesium bicarbonate	$Mg(HCO_3)_2$	37 100
Magnesium carbonate	$MgCO_3$	101
Magnesium chloride	$MgCl_2$	362 000
Magnesium hydroxide	$Mg(OH)_2$	17
Magnesium sulphate	$MgSO_4$	170 000
Sodium bicarbonate	$NaHCO_3$	38 700
Sodium carbonate	Na_2CO_3	61 400
Sodium chloride	$NaCl$	225 000
Sodium hydroxide	$NaOH$	370 000
Sodium sulphate	Na_2SO_4	33 600

$$\text{Mg(OH)}_2 \quad \underset{\text{precipitation}}{\overset{\text{dissolution}}{\rightleftarrows}} \quad \text{Mg}^{++} + 2 \text{ OH}^-$$

and from fundamental thermodynamics the saturated equilibrium
equation for the above reaction is:

$$\frac{(\text{Mg}^{++}) \ (\text{OH}^-)^2}{(\text{Mg(OH)}_2 \ (\text{Solid})} = \text{constant} \tag{1}$$

where ( ) indicate active concentrations.

At saturated equilibrium the concentrations of Mg^{++} and OH^-
in solution are independent of the quantity of solid Mg(OH)_2 in
contact with the solution. Consequently, the term $(\text{Mg(OH)}_2$
solid) in Eq. (1) can be set equal to unity giving the solubility product equation for magnesium hydroxide:

$$(\text{Mg}^{++})(\text{OH}^-)^2 = K_{S(\text{Mg(OH)}_2)} \tag{2}$$

where $K_{S(\text{Mg(OH)}_2)}$ is defined as the thermodynamic solubility
product constant for magnesium hydroxide.

Similarly for calcium carbonate, at saturation the solubility
product equation is:

$$(\text{Ca}^{++})(\text{CO}_3^=) = K_{S(\text{CaCO}_3)} \tag{3}$$

where $K_{S(\text{CaCO}_3)}$ is the thermodynamic solubility product for
CaCO_3.

The values, K_S in Eqs. (2 and 3) are significantly affected
by temperature.

In general the concentration of solute (dissolved solid in
the liquid) is independent of the mass of undissolved solid in
contact with the liquid. However, if the solid occurs as micro-
scopic crystalline dispersed particles, the solid surface in
contact with the liquid is inordinately large, and a state of

metastable solubility equilibrium may exist, appreciably altering the solubility of the solid.

The rate at which solids dissolve and precipitate from solution is of practical importance. In general, compared with the
time required for the establishment of equilibrium between aqueous species (usually of the order of mille seconds), the establishment of the equilibrium between the solute and the solid is
a slow process and may take from a few minutes to years. The
time to reach solubility equilibrium is a complex function of
such factors as 'the degree of under-or oversaturation', mixing
conditions in the water, charge on the crystal seed, availabil-
ity of crystal growth sites and thermodynamic properties of the
solid in question. This is a kinetic problem, and although important, will not be considered in this monograph which will
only deal with equilibrium conditions.

Effect of Temperature on the Solubility Product of $CaCO_3$

Frear and Johnston (1929) determined *ratios of solubility* of
calcium carbonate in the range 0°C to 50°C compared with the
solubility at 25°C. Larson and Buswall (1942) using the data
of Frear and Johnston and other investigations determined values
of the solubility product, for unit activity, in the temperature
range 0°C to 80°C. A graphical plot of their data yields the
following relationship:

$$pK_S = 0,01183t + 8,03 \qquad (4)$$

where t is deg.C.

From Eq. (4) as temperature increases, pK_S for calcium carbonate increases, that is, K_S decreases.

Effect of Temperature on the Solubility Product for $Mg(OH)_2$

Magnesium hydroxide often occurs in water treatment processes
as a finely divided barely crystalline metastable precipitate
(Hamer, 1961). In this state it may have a much greater

solubility in water than in a stable crystalline form (Ryznar *et al*, 1946). According to Hamer this factor is partly responsible for the wide spread of reported values on its solubility. To develop a conditioning diagram (Chapter 5) it is necessary to accept some reasonable set of values relating the solubility of magnesium hydroxide with temperature. From data by Travers and Nouvel (1929), Carlson (1952) and Hamer (1961), one can relate $pK_{Mg(OH)_2}$ with temperature (see Figure 1). There is clearly a linear relationship between $pK_{Mg(OH)_2}$ and temperature in the range 0°C to 80°C, which can be expressed by the following relationship:

$$pK_{Mg(OH)_2} = 0,0175 \ T + 9,97 \tag{5}$$

where T is temperature in deg. C and activity is unity.

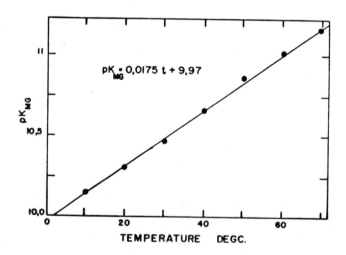

Figure 1. Variation of solubility product for $Mg(OH)_2$ with temperature.

Effect of Ionic Strength

It is of practical convenience to express the solubility product
equations (for example, Eq. (3)) in terms of molar concentrations instead of active concentrations. This is achieved by
adjusting the thermodynamic solubility product to include the
effects of ionic strength. Rewriting Eq. (3) in terms of molar
concentrations and activity coefficients:

$$f_D[Ca^{++}] \ f_D[CO_3^=] = K_S \tag{6}$$

Transferring the activity coefficients in Eq. (6) to the right
hand side of the equation:

$$[Ca^{++}] \ [CO_3^=] = K_S/f_D^2 \tag{7}$$

Incorporating the activity factors in the thermodynamic product,
K_S, gives the effective solubility product K_S' which now includes
the effects of both temperature and ionic strength:

$$[Ca^{++}] \ [CO_3^=] = K_S' \tag{8}$$

The activity coefficient f_D decreases with increase in ionic
strength (see Chapter 2), hence from Eq. (7) the greater the
ionic strength of a liquid the more soluble will be the dissolving substance.

Common Ion Effect

If Eq. (8) is used to determine the solubility of $CaCO_3$ in a
particular water, the initial condition of the water should be
taken into account. For example, the concentration of $CaCO_3$
dissolving into pure CO_2 free water, (determined from Eq. (8))
exceeds the $CaCO_3$ concentration dissolving into a CO_2 free
water containing calcium ions. This phenomenon is due to the
common ion effect (in this case - calcium). Suppose a water
contains both calcium and carbonate ions in such concentrations

that the solubility product constant is not exceeded, then the
addition of a common ion, either Ca^{++} (by adding say $CaCl_2$) or
$CO_3^=$ (by adding say Na_2CO_3), can cause the $CaCO_3$ solubility product to be exceeded and $CaCO_3$ precipitation.

Calcium Carbonate Saturation

If Total Alkalinity, pH and calcium concentration of a water are
measured, it is possible to calculate whether or not a water is
saturated with respect to calcium carbonate, i.e. whether or not
the solubility product equation for $CaCO_3$ is satisfied.

Where the product of active concentration of calcium and carbonate ions exceeds the solubility product constant, the water
is oversaturated with respect to $CaCO_3$ and $CaCO_3$ will precipitate. Such a water will exhibit a tendency to be scale forming.
Where the solubility product is not exceeded the water is undersaturated with respect to $CaCO_3$ and such a water will tend to be
scale dissolving.

Historically the most important attempt to establish whether
a water is scale forming or scale dissolving is due to Langelier
(1936). From theoretical considerations he developed a 'saturation index' which, by its sign, establishes the state of a water
with respect to over- or undersaturation of $CaCO_3$. The Saturation (or Langelier) Index, and variations of it, has dominated
water stabilization practice since its inception, despite re-
servations expressed at various times.

The Langelier Index is best introduced from the following
reasoning: the carbonic species concentrations in a water can
be established by measuring Alkalinity and pH (see Chapter 3).
Knowing the concentration of $CO_3^=$ ions, a theoretical Ca^{++} concentration can be calculated from the $CaCO_3$ saturation equation,
Eq. (8), to give the concentration of Ca^{++} which will just satisfy
solubility equilibrium at the observed pH and Alkalinity. If
the actual Ca^{++} concentration is less than the theoretical concentration the water is undersaturated, if greater the water is

oversaturated and $CaCO_3$ will precipitate. Instead of using the $CaCO_3$ concentration as the measure of over- or undersaturation, Langelier calculated the theoretical pH of the water (using the measured Alkalinity and calcium concentration) at which the water would be just saturated with respect to $CaCO_3$ and called this theoretical pH value the saturated pH value, pH_S. The Saturation Index is defined as $(pH_{actual} - pH_S)$ and is negative for undersaturation and positive for oversaturation.

Langelier emphasized that the Index must be considered only as a qualitative measure of over- or undersaturation. This point has not always been appreciated, and waters are often conditioned, say, to give a positive Index of +0,2 and this Index value interpreted as if it is directly related to the mass of $CaCO_3$ expected to deposit in the pipe system carrying the water. However, a positive Index guarantees neither (1) stability of the Index, nor (2) the mass of potential precipitatate.

(1) The buffering capacity of the water governs the resistance to pH change. However, the buffer capacity changes in a complex manner with the pH (see Chapter 3, Figure 8), so that in certain pH regions appreciable changes occur from minor external influences. This is particularly true for waters conditioned to pH values in the region 8,3. At this pH value there is virtually no buffering capacity so that pH fluctuations are virtually inevitable and such waters, conditioned to a slightly positive Index, may exhibit fluctuations of over- and undersaturation, i.e. positive and negative Indices. In this situation the Index is of doubtful merit as a parameter in controlling the scale depositing characteristics of the water.

(2) The total pH change which occurs when all the $CaCO_3$ precipitates from an oversaturated water does not equal the Index value. The reason for this is that all the

parameters pH, Ca^{++} and Alkalinity change with precipitation and the actual change in pH will vary depending on
the initial pH, Alkalinity and calcium concentration. Hence
the Index, and modifications of it, is of little use in
evaluating the mass of potential precipitate.

A further criticism of the Index is the postulation that if a
water is oversaturated the resulting precipitation on the walls
of pipes carrying such waters would provide a film protecting
the metal from corrosion. Work by Stumm, (1962), has shown
that this postulation is not generally valid and that the formation of corrosion protective films appears to be related to
both oversaturation and buffer capacity in the water. Stumm
showed that oversaturated waters with low buffer capacities
give rise to deposits which are tubercular in appearance and do
not prevent corrosion; whereas oversaturated water with high
buffer capacity gives rise to dense uniform corrosion preventa-
tive films.

Evidence supporting Stumm's postulation regarding the effect
of buffering capacity is found by comparing the behaviour of
treated waters from New York and Cape Town. In Table 2 the constitution of the treated waters from these two cities is listed.

Comparing the two waters, the Saturation Indices are not significantly different and the theoretical $CaCO_3$ deposition poten-
tials (see Chapter 5) are identical. Cape Town's calcium concentration and Alkalinity is in fact higher than New York's. On
the basis of these observations one would tend to conclude that
Cape Town's water should, if anything, be more stable than New
York's. However, according to Langelier (1936), New York's
water was non-corrosive but in Cape Town tubercular corrosion
is evident. The main difference between the two waters is in
the pH and buffering capacity, both of which are higher in New
York than in Cape Town. In particular, the buffering capacity

Table 2

	pH	Ca ppm as $CaCO_3$	Alk. ppm as $CaCO_3$	Saturation Index***	C_T ppm as $CaCO_3$	$CaCO_3$ precipitation potential ppm	Buffer capacity (moles H^+/unit) $.10^{-5}$
New York City*	9,7	25	18	+ 0,3†	15	2	1,5
Cape Town City**	9,0	37	32	+ 0,2	28	2	0,5

*   Tests reported by Langelier (1936)

**   From data supplied by Morrison (1970)

***   Calculated from Langelier's equations, Eqs. (9 and 17)

†   Reported incorrectly as + 0,4 in Langelier's paper, 1936.

of New York's water is three times that of Cape Town. This
latter observation is possibly the crucial reason for the difference in corrosive behaviour of the two waters which tends to
support Stumm's postulation. This comparison emphasizes the
importance of considering both oversaturation and buffer capacity as the criteria for corrosion protection by scale formation
particularly in soft waters.

Viewed in retrospect the success attained by the use of
Langelier's Index is due possibly to most treated waters having
a relatively high total carbonic species concentration and as
such the buffer capacity is sufficient for uniform film formation even in the region of low buffer capacity, pH 8 to 9, provided the water is oversaturated.

The importance and prestige which Langelier's Index has
achieved with time is sufficient proof of the favourable consequences from its application. However, as better understanding
is accumulating on the complex nature of the chemistry of water,
it is essential that the Index be reassessed and its advantages
and limitations defined more specifically. To do this requires
a thorough appreciation of Langelier's concepts, and only then
can a critical assessment be made of some of the ideas presented
in Chapter 5, where the general problem of water conditioning
will be discussed.

Langelier's Theory of the Saturation Index

Langelier used the term 'pH value at saturation', pH_S, to mean
that theoretical pH value at which a water of a given Alkalinity
and calcium content would be just saturated with $CaCO_3$. The
saturation state of a water is given by the Saturation Index
(S.I.) which is defined as the difference between pH of the
water and the theoretical pH for $CaCO_3$ saturation, pH_S, i.e.

$$S.I. = pH_{actual} - pH_S \tag{9}$$

If the Saturation Index is zero the water is just saturated with $CaCO_3$. If positive, the water is oversaturated, and if negative, is undersaturated with respect to $CaCO_3$. Langelier does not define whether the 'pH values' in Eq. (9) are the 'p' values of the active hydrogen ion concentration or the molar hydrogen ion concentration. However, noting that pH_{actual} is inevitably a pH meter reading, the corresponding pH_S value should also be in terms of the active hydrogen ion concentration.

An equation for pH_S is developed below from fundamental ionic and solubility equilibria reactions following the same procedure proposed by Langelier:

$$Alkalinity = [A] = [HCO_3^-] + 2[CO_3^=] + [OH^-] - [H^+]$$

where $[H^+]$ is the molar hydrogen ion concentration derived from the measured pH in the water.

Solving for $[CO_3^=]$:

$$[CO_3^=] = ([A] - [HCO_3^-] - [OH^-] + [H^+])/2 \qquad (10)$$

Substituting for $[HCO_3^-]$ from Eq. (30), Chapter 3, into Eq. (10):

$$[CO_3^=] = \frac{K_2' ([A] - [OH^-] + [H^+])}{[H^+](1 + 2K_2'/[H^+])} \qquad (11)$$

Up to this stage evaluation of $[CO_3^=]$ is from a straightforward application of ionic equilibria. Langelier now, in effect, asks the following question: if $[Ca^{++}]$ is known, what should be the pH of the particular water to satisfy $CaCO_3$ saturation? This pH value is found indirectly through the $[CO_3^=]$ value at saturation, $[CO_3^=]_S$:

From Eq. (8):

$$[CO_3^=]_S = K_S'/[Ca^{++}] \qquad (12)$$

but for a fixed Alkalinity in a water $[CO_3^=]$ is a function only of pH (see Eq. (11)), i.e. $[H^+]$ must change to give $[CO_3^=]_S$ on the lefthand side of Eq. (11). Hence, substituting $[CO_3^=]_S$ for $[CO_3^=]$ one may solve for the required $[H^+]$ at saturation $[H^+]_S$. Substituting for $[CO_3^=]_S$ from Eq. (8) into Eq. (11):

$$\frac{[Ca^{++}]K_2' \, ([A] - [OH^-]_S + [H^+]_S)}{[H^+]_S \, (1 + 2K_2'/[H^+]_S)} = K_S' \qquad (13)$$

where $[H^+]_S$ and $[OH^-]_S$ are the molar hydrogen and hydroxide ion concentrations respectively at saturation for the measured Ca^{++} and Alkalinity values.

Taking the logarithm of Eq. (13) and simplifying, gives:

$$-\log[H^+]_S = -\log K_2' + \log K_S' - \log [Ca^{++}] - \log ([A] - [OH^-]_S$$
$$+ [H^+]_S) + \log (1 + 2K_2'/[H^+]_S) \qquad (14)$$

Noting that $pH_S = -\log (f_m[H^+]_S)$, Eq. (14) is rewritten as below:

$$-\log (f_m[H^+]) = -\log K_2' + \log K_S' - \log [Ca^{++}] - \log ([A]$$
$$- [OH^-]_S + [H^+]_S) + \log(1 + 2K_2'/[H^+]_S)$$
$$- \log f_m \qquad (15)$$

Rewriting this equation using the pX notation gives the general equation for the pH at saturation, pH_S, i.e.

$$pH_S = pK_2' - pK_S' + p[Ca^{++}] + p([A] - [OH^-]_S + [H^+]_S)$$
$$+ \log (1 + 2K_2'/[H^+]_S) - \log f_m \qquad (16)$$

The general equation for pH_S proposed by Langelier is obtained by rewriting in the pX notation, i.e.

$$p[H]_S = pK_2' - pK_S' + p[Ca^{++}] + p([A] - [OH^-]_S$$

$$+ \log (1 + 2K_2'/[H^+]_S) \qquad (17)$$

Comparing Eqs. (16 and 17), using Langelier's formula for pH_S, one is ignoring the effect of ionic strength on pH, i.e. an error of magnitude $\log f_m$ is introduced. Values of $\log f_m$ for a number of TDS values, calculated from Langelier's formula, Eq. (28) Chapter 3, and the Davies equation, Eq. (96) Chapter 2, for activity coefficients, are given in Table 3. The error in using Langelier's original equation, Eq. (17), is only of any significance for TDS greater than 500 ppm, however, because $(+ \log f_m)$ has a negative value, Langelier's equation results in overestimating the Saturation Index.

Table 3.

T.D.S.(ppm)	50	200	300	500	1000
$+ \log f_m$	$-0,02$	$-0,03$	$-0,04$	$-0,05$	$-0,07$

The hydrogen ion concentration at saturation, $[H^+]_S$, occurs implicitly in the Eq. (16). Hence solution for pH_S is only possible by iteration.

For most waters a number of simplifying assumptions can be made in specific pH ranges:

(i) For $9,5 < pH < 10,3$, hydrogen and hydroxide ion concentrations are negligible in the term $([A] - [OH^-]_S + [H^+]_S)$ and Eq. (16) reduces to:

$$pH_S = pK_2' - pK_S' + p[Ca] + p[A] + \log (1 + 2K_2'/[H^+]_S) - \log f_m$$

$$(18)$$

This equation also has to be solved by iterative techniques due to the presence of $[H^+]_S$ in the righthand side of the equation.

(ii) For $6,5 < pH < 9,5$ the term $2K_2'/[H^+]_S$ is negligible and Eq. (18) reduces to:

$$pH_S = pK_2' - pK_S' + p[Ca] + p[A] - \log f_m \qquad (19)$$

This equation provides a direct solution for pH_S. Most natural waters lie in this range so that Eq. (19) is of wide utility.

(iii) For $pH < 6,5$ the terms $2K_2'/[H^+]_S$ and $[OH-]_S$ are negligible and Eq. (15) approximates to:

$$pH_S = pK_2' - pK_S' + p[Ca] + p([A] + [H^+]_S) - \log f_m \qquad (20)$$

and again pH_S has to be solved by an iterative procedure.

In Chapter 5, which deals with graphical solutions to conditioning problems, a simple graphical procedure in which no approximations are made, will be presented to calculate the Index.

It will be shown that where the pH in a water is greater than the carbonate equivalence point (i.e. greater than pH about 10,4) the Index reverses and a positive Index indicates undersaturation and negative oversaturation.

However, more important is that a simple graphical method is given which defines the mass of $CaCO_3$ which will either precipitate or be dissolved from an over- or undersaturated water to achieve saturated equilibrium, and thus gives a quantitative measure of the degree of over- or undersaturation. Furthermore, simple graphical methods are proposed to give a quick accurate estimate of the buffer capacity, β, for any measured pair of values for Alkalinity and pH.

References

Carlson,E.T., Peppler,R.B. and Wells,L.S. 1953 "System Magnesia, Silica, Water at Elevated Temperature and Pressures", J.Res.Nat.Bur.Stand., 51, 179.

Frear,G.L. and Johnston,J. 1929 "The Solubility of Calcium Carbonate in Certain Aqueous Solutions at 25°C", J.A.Chem.Soc., 51, 2082.

Hamer,P, Jackson, J. and Thurston,E.F. 1961 *Industrial Water Treatment Practice*, Butterworth, London.

Harned, H.S. and Scholes,S.R. 1941 "The Ionization Constant of HCO_3^-", J.Am.Chem.Soc., 63, 1706.

Langelier,W.F. 1936 "The Analytical Control of Anti-Corrosion Water Treatment", Jour. A.W.W.A., 28, 1500.

Larson,T.E. and Buswell,A.M. 1942 "Calcium Carbonate Saturation Index and Alkalinity Interpretation", Jour. A.W.W.A., 34, 1667

Nordell, E. 1961 *Water Treatment for Industrial and Other Uses*, Second Edition, Rheinhold Publ.Corp., New York.

Ryznar,J.W., Green, J. and Winterstein,M.G. 1946 "Determination of the pH of Saturation of Magnesium Hydroxide". Indus. Eng.Chem., 38, 1057.

Stumm,W. 1960 "Investigation on the Corrosive Behaviour of Waters". A.S.C.E., 86,2657.

Travers,A. and Nouvel,T. 1929 "The Solubility of Magnesium Hydroxide at Elevated Temperatures", C.R.Acad. Sci.Paris, 188, 499.

Chapter 5

WATER CONDITIONING

In water conditioning for general public use one is concerned with adjusting pH, Alkalinity, Ca^{++} and Mg^{++} to values which will minimize the possibility of corrosion problems in pipe lines; provide a water not too hard for domestic use and which will not quickly encrust warm water fittings. A usual requirement is that the treated water is saturated or slightly oversaturated with respect to $CaCO_3$ but that the Ca^{++} concentration is not too high.

Problems in conditioning can be divided into 3 general types, involving:

1. Aqueous (or single) Phase Equilibrium

Problems involving only this phase of equilibrium usually arise in conditioning waters initially undersaturated with respect to $CaCO_3$. It is necessary to adjust pH, Alkalinity and Ca^{++} so that the water is saturated, or slightly oversaturated, with respect to $CaCO_3$ and has an appropriate pH. (If the Ca^{++} content, pH and degree of saturation desired are specified, the Alkalinity corresponding to these requirements is fixed and can be calculated).

$CaCO_3$ precipitation from a slightly oversaturated, unseeded water is a slow process compared with the rate at which equilibrium is achieved between the dissolved species. Consequently, for the purposes of conditioning, calculations may be performed on the assumption that calcium and carbonate remain in the ionized form even though the water is oversaturated. The solubility product of $CaCO_3$ does not therefore enter into the calculations. This assumption is valid where the degree of oversaturation is not too great and the water unseeded.

A further assumption inherent in these calculations is that there is no CO_2 exchange with the air. This assumption is usually valid when treating surface waters but is not necessarily valid when treating underground water which may contain high concentrations of CO_2 in which event CO_2 exchange between the water and air will occur. Where CO_2 exchange takes place, conditioning problems may involve two or three phases and are considered in (3) below.

Aqueous or single phase equilibrium problems therefore assume (a) all chemicals are infinitely soluble and occur only in the dissolved phase (in particular $CaCO_3$, $Mg(OH)_2$ and CO_2); and (b) no CO_2 gas exchange takes place between the water and air.

Chemical conditioning involving only single aqueous phase equilibrium has wide application, particularly in the final stabilization process of water treatment.

2. Two Phase or Solid Aqueous Phase Equilibrium

Conditioning problems involving solid and aqueous phases of equilibria are important where either a water is 'hard' i.e. contains high concentrations of Ca^{++} and Mg^{++} ions, or where a water, undersaturated with respect to say $CaCO_3$, comes into contact with a source of solid $CaCO_3$.

Where a water is hard, chemical treatment may be required to reduce the Ca^{++} and Mg^{++} concentrations to acceptable limits. In practice this is achieved by precipitating these cations from solution as solid $CaCO_3$ and $Mg(OH)_2$ respectively. In this process two phases are involved, the aqueous phase and the solid phase in which the solubilities of $CaCO_3$ and $Mg(OH)_2$ are involved.

Solution of conditioning problems involving aqueous and solid phases again assumes (as in (1) above) that there is no CO_2 exchange between the water and air.

Problems involving aqueous-solid phase equilibria are of widespread occurrence throughout the hinterland of South Africa. The most important example is the softening of the water supply to the Rand by the Rand Water Board.

3. Three Phase or Gas-Aqueous-Solid Phase Equilibrium

Problems involving equilibrium between three phases occur where a water is in contact with air or a gas containing CO_2. Carbon dioxide either enters or leaves the water depending on the partial pressures of CO_2 in the water and in the gas. Carbon dioxide exchange affects both Acidity and pH of the water and could result in either $CaCO_3$ over- or undersaturation depending on whether CO_2 is expelled or absorbed by the water. Three phase equilibrium then involves equilibria relations between the ionic, $CaCO_3$-solid and CO_2 gas phases.

The rate of gas transfer across the air-liquid interface depends on the CO_2 partial pressures, water surface exposed to the gas and the degree of turbulence. Total CO_2 transfer will be dependent on the time of contact under the conditions above.

In water treatment practice the effect of CO_2 in the air on water stored in service reservoirs will probably be minor due to the short holding times. Carbon dioxide exchange may, however, be of great importance in determining the final condition of water from boreholes having a high CO_2 content in which CO_2 is removed by some stripping process.

In rivers, particularly those running over calcareous and dolomitic bed material, equilibrium is eventually established between the solubilities of Ca^{++} and Mg^{++} with the carbonic species in solution, and the carbonic system in solution with the CO_2 partial pressure of the air. Treated sewage effluents in the receiving streams may exchange CO_2 with the air, changing the pH, Acidity and perhaps Ca^{++} concentrations.

In general, the quality of a water, particularly with regard to the carbonic system and Ca^{++} and Mg^{++} concentrations, could be significantly affected by CO_2 exchange and the nature of the bedrock over which it flows. Due to the time factor involved in gas transfer and solubility reactions, equilibrium is often not

achieved. Using equilibria relationships one can predict only
the eventual equilibrium conditions. However, this information
is valuable in defining the expected direction of change in the
state of a water. Such information is often sufficient in making
decisions. For example in storage dams receiving hard waters,
CO_2 exchange with the air or CO_2 removal due to algal photosynthesis may raise the pH in the surface water sufficiently to
cause $CaCO_3$ precipitation. If such a dam is used as a source for
domestic water the abstraction point may be preferable from the
surface layers where hardness has been reduced. Abstraction from
the bottom layers may result in having a water with increased
hardness caused by resolution of $CaCO_3$ due to the lowering of the
pH when CO_2 is produced in biological oxidation of organic matter
in the lower depths.

For any particular conditioning problem inevitably a number
of feasible solutions exist and choice of the 'best solution'
requires that alternative treatment procedures be considered.

Numerical methods of solution to conditioning problems involving single, two and three phase systems are not practical for
everyday use due to the complex inter-relationships between the
variables defining equilibrium conditions. However, practical
methods of solution are possible using graphical plots (or conditioning diagrams) to define equilibria inter-relationships between basic variables or functions of these variables (e.g.
Alkalinity, Acidity, C_T).

Before developing conditioning diagrams and graphical solutions to water treatment problems it is necessary to list some of
the chemicals commonly used in water conditioning; to compare the
forms by which their concentrations are usually expressed and to
examine the effects which addition of these chemicals has on the
parameters Alkalinity, Acidity and total carbonic species concentration.

Molar and Equivalent Concentrations and Concentration in ppm
Expressed as $CaCO_3$

Chemicals used in water conditioning inevitable include an ion or
compound which is common to at least one of the parameters in the
calcium-magnesium-carbonic system, i.e.

Dosing Chemical	Common Parameter
$Ca(OH)_2$	Ca^{++} and OH^-
Na_2CO_3	$CO_3^=$
$NaHCO_3$	HCO_3^-
CO_2	$H_2CO_3^*$
Strong base	OH^-
Strong acid	H^+
$CaCl_2$	Ca^{++}
$Mg(OH)_2$ precipitation	Mg^{++} and OH^-

Concentrations of chemicals are usually expressed in one of
three forms in water treatment: molar concentrations, equivalent
concentrations or 'concentration in ppm expressed as $CaCO_3$'. In
this monograph to distinguish between these the following format
is used throughout:

[ ] indicates molar concentrations
{ } indicates equivalent concentrations
'no brackets' indicates a concentration in ppm
    expressed as $CaCO_3$

Molar concentration is defined as the number of gram moles of
a chemical dissolved in a litre of solution, i.e.

$$[\ ] = \frac{\text{mass in grams of a chemical/} \ell \text{ of solution}}{\text{molecular weight of the chemical}} \qquad (1)$$

Equivalent concentration is defined as the number of gram
equivalents of a chemical in a litre of solution, i.e.

$$\{ \; \} = \frac{\text{mass in grams of a chemical}/\ell \text{ of solution}}{\text{equivalent weight of the chemical}} \qquad (2)$$

Now, equivalent weight = molecular weight/valency. Substituting this expression into Eq. (2):

$$\{ \; \} = \frac{\text{mass in grams of chemical}/\ell \text{ of solution}}{\text{molecular weight of chemical/valency}}$$

$$= \frac{\text{mass in grams of chemical}/\ell \text{ of solution}}{\text{molecular weight of chemical}} \; x \; \text{valency}$$

$$= [\;] \; x \; \text{valency} \qquad (3)$$

Conversion of concentrations expressed in molar form to the equivalence form is explained through the example below:

Example

Convert the equations for Alkalinity, Acidity and total carbonic species concentration from the molar form to the equivalence form.

By definition:

$$[\text{Alkalinity}] = [HCO_3^-] + 2[CO_3^=] + [OH^-] - [H^+]$$

where [Alkalinity] is the moles of H^+ required to convert one litre of solution to an equivalent carbonic acid solution.

The terms $[HCO_3^-]$, $[OH^-]$ and $[H^+]$ are all monovalent, and from Eq. (3) their concentrations in the molar form equals the concentration in the equivalence form.

For the $[CO_3^=]$ term:

$$2 \; x \; [CO_3^=] = \{CO_3^=\}$$

Thus the formula for Alkalinity in terms of equivalent concentrations is:

$$\{\text{Alkalinity}\} = \{HCO_3^-\} + \{CO_3^=\} + \{OH^-\} - \{H^+\} \qquad (4)$$

where {Alkalinity} is the equivalents of H^+ required to convert one litre of solution to an equivalent carbonic acid solution.

Similarly for Acidity the molar form

$$[\text{Acidity}] = 2[H_2CO_3^*] + [HCO_3^-] + [H^+] - [OH^-]$$

reduces to the equivalent form

$$\{\text{Acidity}\} = \{H_2CO_3^*\} + \{HCO_3^-\} + \{H^+\} - \{OH^-\} \tag{5}$$

and for total carbonic species, the molar form:

$$[C_T] = [H_2CO_3^*] + [HCO_3^-] + CO_3^=]$$

reduces to the equivalent form

$$\{C_T\} = \{H_2CO_3^*\}/2 + \{HCO_3^-\} + \{CO_3^=\}/2 \tag{6}$$

In water treatment, the most common form of expressing chemical concentrations is as '*ppm expressed as* $CaCO_3$'. In Table 1 are listed factors for converting the concentrations of some of the common chemicals from ppm to ppm expressed as $CaCO_3$. The method of conversion is in each case similar to the following example:

Example

X ppm (i.e. mg/ℓ) of $Ca(OH)_2$ is added to water. What concentrations of Ca^{++} and OH^-, expressed as 'ppm as $CaCO_3$' have been added to the water?

m eq/ℓ of $Ca(OH)_2$ added = X/E.W. of $Ca(OH)_2$ = X/37

i.e.

m eq/ℓ of Ca^{++} added = X/37

and

m eq/ℓ of OH^- added = X/37

i.e.

concentration Ca^{++} added expressed as $CaCO_3$ equals the concentration of OH^- added, both expressed as $CaCO_3$

$$= \frac{X}{37} * \text{E.W. } CaCO_3$$

$$= \frac{X}{37} * \frac{100}{2}$$

$$= X * \frac{50}{37}$$

Table 1.

Chemical with Concentration X mg/ℓ	Concentration of chemical as m eq/ℓ	Parameter	Concentration of parameter as ppm $CaCO_3$
$Ca(OH)_2$	X/37	Ca^{++}	X*50/37
		OH^-	X*50/37
CO_2	X/22	CO_2	X*50/22
Na_2CO_3	X/53	$CO_3^=$	X*50/53
$NaHCO_3$	X/84	HCO_3^-	X*50/84
$NaOH$	X/40	OH^-	X*50/40
$CaCl_2$	X/58	Ca^{++}	X*50/58
H_2SO_4	X/49	H^+	X*50/49
$CaCO_3$	X/50	Ca^{++}	X
		CO_3^{--}	X

Equations for Alkalinity, Acidity and total carbonic species concentration, Eqs. (4 to 6), can now be written with concentrations expressed in ppm a $CaCO_3$, i.e.

$$\text{Alkalinity} = HCO_3^- + CO_3^= + OH^- - H^+ \tag{7}$$

$$\text{Acidity} = H_2CO_3^* + HCO_3^- + H^+ - OH^- \tag{8}$$

$$C_T = H_2CO_3^*/2 + HCO_3^- + CO_3^=/2 \tag{9}$$

Furthermore, the fundamental equilibria equations for the calcium-carbonic system, Eqs. (7 to 10) in Chapter 3 and Eq. (8) in Chapter 4, can be rewritten with species concentrations expressed in ppm as $CaCO_3$ by modifying the equilibrium constants as follows:

From Eq. (7), Chapter 3:

$$\frac{H^+/5*10^4 \times HCO_3^-/5*10^4}{H_2CO_3^*/10^5} = K_1'$$

i.e.

$$\frac{H^+ . HCO_3^-}{H_2CO_3^*} = K_1'*2,5*10^4 = K_{c1} \tag{10}$$

From Eq. (9), Chapter 3:

$$\frac{H^+/5*10^4 \times CO_3^=/10^5}{HCO_3^-/5*10^4} = K_2'$$

i.e.

$$\frac{H^+ . CO_3^=}{HCO_3^-} = K_2'*10^5 = K_{c2}' \tag{11}$$

and from Eq. (10) Chapter 3:

$$H^+/5*10^4 \times OH^-/5*10^4 = K_W'$$

i.e.

$$H^+ \cdot OH^- = K_W' *25*10^8 = K_{cW}' \tag{12}$$

and from Eq. (8), Chapter 4:

$$CO_3^=/10^5 \times Ca^{++}/10^5 = K_S'$$

i.e.

$$Ca^{++} \cdot CO_3^= = K_S'*10^{10} = K_{cS}' \tag{13}$$

Changes in Alkalinity, Acidity and C_T with Chemical Dosing

Alkalinity, Acidity and total carbonic species concentration in
a water change in a straightforward manner with chemical dosing.
These parameters are of prime importance in developing graphical
aids for solution to water treatment problems. Changes in pH,
however, are complex and can only be calculated using equilibria
considerations.

The respective changes in the parameters Alkalinity, Acidity
and total carbonic species concentration are expressed as follows:

Alkalinity change = Alkalinity added or removed   (14)

Acidity change   = Acidity added or removed   (15)

Total carbonic   = carbonic species added or   (16)
species change   removed

From the definitions of Alkalinity, Acidity and total carbonic species concentration, (Chapter 3, page 99), Eqs. (14 to
16) are more explicitly defined as follows:

$$\text{Alkalinity}_{added} = CO_3^={}_{added} + HCO_3^-{}_{added} + OH^-{}_{added} - H^+{}_{added} \tag{17}$$

$$\text{Acidity}_{added} = H_2CO_3^*{}_{added}\ HCO_3^-{}_{added} + H^+{}_{added} - OH^-{}_{added} \tag{18}$$

Carbonic species = $H_2CO_3^*$ $_{added}/2$ + HCO_3^- $_{added}$ + $CO_3^=$ $_{added}/2$ (19)
added

and comparing Eqs. (14 to 16) with Eqs. (17 to 19) gives:

Alkalinity = $CO_3^=$ $_{added}$ + HCO_3^- $_{added}$ + OH^- $_{added}$ − H^+ $_{added}$ (20)
change

Acidity = $H_2CO_3^*$ $_{added}$ + HCO_3^- $_{added}$ + H^+ $_{added}$ − OH^- $_{added}$ (21)
change

Carbonic = $H_2CO_3^*$ $_{added}/2$ + HCO_3^- $_{added}$ + $CO_3^=$ $_{added}/2$ (22)
species
change

In Table 2 a series of examples are calculated for changes in Alkalinity, Acidity and C_T (using Eqs. (20 to 22)) with addition of various chemicals to a water.

Table 2.

X ppm of chemical added (expressed as $CaCO_3$)	Alkalinity Change (ppm as $CaCO_3$)	Acidity Change (ppm as $CaCO_3$)	C_T Change (ppm as $CaCO_3$)
$Ca(OH)_2$	+ X	− X	0
NaOH	+ X	− X	0
HCl	− X	+ X	0
$NaHCO_3$	+ X	+ X	+ X
Na_2CO_3	+ X	0	+ X/2
CO_2	0	+ X	+ X/2
$CaCO_3$ (added)	+ X	0	+ X/2
$CaCO_3$ (precipitated)	− X	0	− X/2
$CaCl_2$	0	0	0

Although Eqs. (14 to 22) can be accepted intuitively from the definitions of Alkalinity, Acidity and C_T, a fundamental approach is to prove the relationships rigorously from basic chemical relationships.

(1) <u>Effect of HCO_3^- Addition on (a) Alkalinity and (b) Acidity</u>

   (a) <u>Alkalinity</u>

      X ppm (expressed as $CaCO_3$) of $NaHCO_3$ is added to a water. For electroneutrality the sum of the changes in positive ions and negative ions are equal, i.e.

$$Na^+_{added} + \Delta H^+ = \Delta HCO_3^- + \Delta CO_3^= + \Delta OH^-$$

i.e.

$$Na^+_{added} = \Delta HCO_3^- + \Delta CO_3^= + \Delta OH^- - \Delta H^+$$

$$= \Delta \text{ Alkalinity}$$

Also $Na^+_{added} = HCO_3^-{}_{added}$

i.e.

$$HCO_3^-{}_{added} = \Delta \text{ Alkalinity}$$

   (b) <u>Acidity</u>

   (i) Mass balance equation with concentrations expressed in moles is:

$$[HCO_3^-]_{added} = \Delta C_T$$

$$= \Delta[H_2CO_3^*] + \Delta[HCO_3^-] + \Delta[CO_3^=]$$

Rewriting this equation with concentrations in the equivalent form:

$$\{HCO_3^-\}_{added} = \Delta\{H_2CO_3^*\}/2 + \Delta\{HCO_3^-\} + \Delta\{CO_3^=\}/2$$

and with concentrations expressed as ppm $CaCO_3$:

$$HCO_{3\ added}^- = \Delta H_2CO_3^* /2 + \Delta HCO_3^- + \Delta CO_3^=/2 \qquad (23)$$

(ii) Proton balance equation for addition of HCO_3^-:

$$\Delta H^+ + \Delta H_2CO_3^*/2 = \Delta CO_3^=/2 + \Delta OH^-$$

and rearranging this equation:

$$\Delta CO_3^=/2 = \Delta H^+ + \Delta H_2CO_3^*/2 - \Delta OH^- \qquad (24)$$

Substituting Eq. (24) into Eq. (32):

$$HCO_{3\ added}^- = \Delta H_2CO_3^* + \Delta HCO_3^- + \Delta H^+ - \Delta OH^-$$

$$= \Delta\ Acidity$$

Thus, if HCO_3^- is added to water Alkalinity increases and Acidity decreases by an amount exactly equal to the amount of HCO_3^- added, provided all the concentrations are expressed in ppm as $CaCO_3$.

(2) Effect of Adding $CO_3^=$ to a Water on the Parameters (a) Alkalinity and (b) Acidity

   (a) Alkalinity

   If, say, X ppm Na_2CO_3 (expressed as $CaCO_3$) is added to a water, then using the same reasoning as in (1a) above for the condition of electroneutrality:

$$Na^+_{added} = CO^=_{3\ added} = \Delta\ Alkalinity$$

   (b) Acidity

   Proton balance for addition of $CO_3^=$ gives:

$$\Delta H^+ + \Delta H_2CO_3^* + \Delta HCO_3^- = \Delta OH^-$$

i.e.

$$\Delta H_2CO_3^* + \Delta HCO_3^- + \Delta H^+ - \Delta OH^- = 0$$

i.e.

$$\Delta \text{ Acidity} = 0$$

That is to say, that if $CO_3^=$ is added to water then Alkalinity changes by an exactly equal amount (if concentrations are expressed as ppm as $CaCO_3$) while Acidity does not change

(3) Effect of Adding CO_2 to a Water on (a) Alkalinity and (b) Acidity.

(a) Alkalinity

Proton balance equation for CO_2 addition:

$$\Delta H^+ = \Delta HCO_3^- + \Delta CO_3^= + \Delta OH^- \tag{25}$$

i.e.

$$\Delta HCO_3^- + \Delta CO_3^= + \Delta OH^- - \Delta H^+ = 0$$

i.e.

$$\Delta \text{ Alkalinity} = 0$$

Thus, if CO_2 is added to water, Alkalinity does not change.

(b) Acidity

Mass balance equation for CO_2 addition is developed as follows:

$$[CO_2]_{added} = \Delta[H_2CO_3^*] + \Delta[HCO_3^-] + \Delta[CO_3^=]$$

and rewriting this equation in terms of equivalent concentrations:

$$\{CO_2\}_{added} = \Delta\{H_2CO_3^*\} + 2\Delta\{HCO_3^-\} + \Delta\{CO_3^=\}$$

and with concentrations expressed in ppm as $CaCO_3$:

$$CO_{2\ added} = \Delta H_2CO_3^* + 2\Delta HCO_3^- + \Delta CO_3^=$$

i.e.

$$\Delta CO_3^= = CO_{2\ added} - \Delta H_2CO_3^* - 2\Delta HCO_3^- \tag{26}$$

Substituting Eq. (26) into Eq. (25) and simplifying:

$$CO_{2\ added} = \Delta H_2CO_3^* + \Delta HCO_3^- + \Delta H^+ - \Delta OH^-$$

i.e.

$$CO_{2\ added} = \Delta\ Acidity$$

Thus the increase in Acidity exactly equals the concentration of CO_2 added provided concentrations are expressed in ppm as $CaCO_3$.

(4) Effect of Adding a Strong Base, say NaOH, to Water on the Parameters (a) Alkalinity and (b) Acidity

(a) Alkalinity

For electroneutrality of the water the changes in the sum of the positive and negative ions must be equal, i.e.

$$Na^+_{added} + \Delta H^+ = \Delta HCO_3^- + \Delta CO_3^= + \Delta OH^- \tag{27}$$

i.e.

$$Na^+_{added} = \Delta\ Alkalinity$$

or the increase in Alkalinity equals the concentration of NaOH added.

(b) <u>Acidity</u>

A mass balance on the change in carbonic species concentration is:

$$\Delta C_T = 0 , \quad \text{or}$$

$$\Delta HCO_3^- + \Delta H_2CO_3^*/2 + \Delta CO_3^=/2 = 0$$

i.e.

$$\Delta CO_3^= = -2\Delta HCO_3^- - \Delta H_2CO_3^* \tag{28}$$

Substituting Eq. (28) into Eq. (27) and simplifying:

$$Na^+_{added} = -\Delta HCO_3^- - \Delta H_2CO_3^* - \Delta H^+ + \Delta OH^-$$

$$= -\Delta \text{ Acidity}$$

In other words, the decrease in Acidity equals the concentration of strong base added.

ALKALINITY-ACIDITY CONDITIONING DIAGRAM

The single phase model considers only changes in equilibrium between the dissolved carbonic species in the aqueous phase due to chemical dosing. The model assumes that no $CaCO_3$ precipitation occurs from the water and that no CO_2 exchange occurs between the water mass and the gas phase (air) in contact with the water.

In Chapter 3 it was shown that provided at least two of the parameters for the carbonic system are measured, the remaining parameters can be calculated using equilibria relationships. Generally, Alkalinity and pH are the two parameters most conveniently measured and hence are usually used to define the state of a water. Because Acidity changes in a straightforward manner with chemical dosing, even though it cannot be readily measured in practice, it is used as an important ancillary theoretical parameter in conjunction with the measured parameters Alkalinity and pH to solve conditioning problems by equilibrium methods: if

pH and Alkalinity of a water are measured, Acidity can always be
calculated using equilibria equations; if a chemical is now added
to the water the new Alkalinity and Acidity values are easily
calculated by simple arithmetic addition and subtraction using
the Δ Alkalinity and Δ Acidity values for the chemicals listed in
Table 2. From the new values for Alkalinity and Acidity and
using equilibrium equations the new pH is calculated.

These three steps constitute the general pattern of solution
to most types of water conditioning problems. The steps involve
the evaluation of one parameter (say Acidity) given the other two
parameters (Alkalinity and pH). These evaluations include equilibria relationships. The only practical method of solution is
by using graphical aids which usually comprise graphical plots
linking Alkalinity, Acidity, pH and other parameters as the case
may be. The basis of these plots and the procedures of solution
will now be considered.

Choosing Alkalinity and Acidity as cartesian co-ordinates,
any other parameter defining the carbonic system, in particular
pH, can be written in terms of Alkalinity and Acidity by applying
equilibria relationships and plotted in terms of the co-ordinate
diagram (see Figure 1). This plot constitutes an equilibrium
diagram for the carbonic system, that is: the measured values of
any two of the parameters Alkalinity, Acidity and pH plot as a
point in the diagram, the equilibrium value of the third parameter is now defined by the value given at the intersection point
and can be read directly off the diagram. For example, water
with Acidity 10 ppm and Alkalinity 5 ppm (both as $CaCO_3$) plots
at Point 1 in Figure 2, the equilibrium pH for this condition,
obtained by interpolation between the pH lines straddling the
intersection point is pH 6,68.

A family of curves representing values for C_T can also be
plotted in the single phase conditioning diagram using the relationship between the mass parameters, Alkalinity, Acidity and

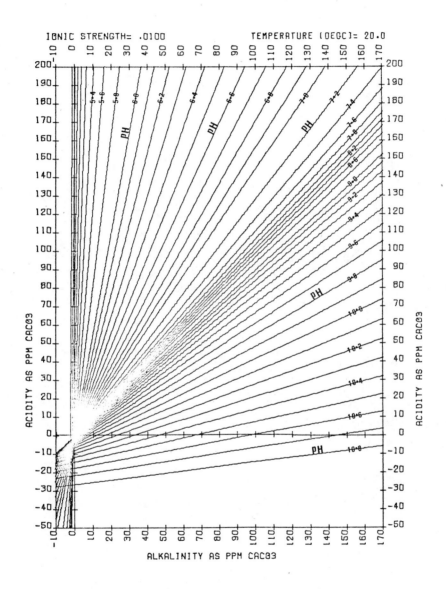

Figure 1. Alkalinity-Acidity-pH conditioning diagram

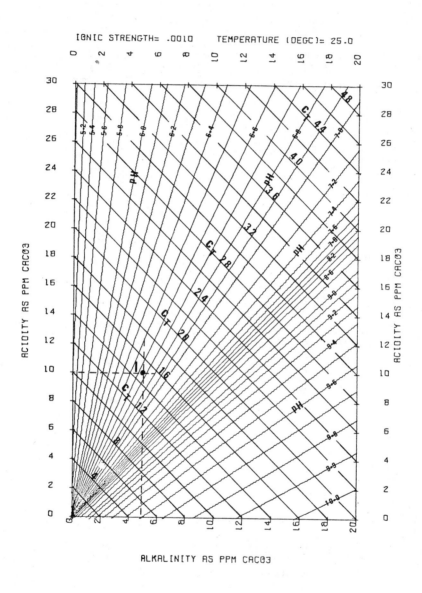

IONIC STRENGTH= .0010   TEMPERATURE (DEGC)= 25.0

ACIDITY AS PPM CACO3

ALKALINITY AS PPM CACO3

Figure 2. Acidity defined by pH and Alkalinity values

C_T, Eq. (9), Chapter 4

Acidity = $2C_T$ - Alkalinity

This equation indicates a linear relationship between the three parameters so that a particular value for C_T plots as a straight line in the diagram and intersects both Acidity and Alkalinity axes at the values $C_T/2$ (see Figure 2).

The theoretical relationships between Alkalinity, Acidity and pH for plotting an equilibrium diagram, such as Figure 1, are developed as follows:

From Eq. (10):

$$H_2CO_3^* = H^+ \cdot HCO_3^-/K_{c1}' \tag{29}$$

Substituting Eq. (29) into Eq. (8):

Acidity = $HCO_3^- + H^+ \cdot HCO_3^-/K_{c1} + H^+ - OH^-$

and substituting Eq. (12) into this equation and solving for HCO_3^-:

$$HCO_3^- = \frac{\text{Acidity} - H^+ + K_{cW}'/H^+}{1 + H^+/K_{c1}'} \tag{30}$$

From Eq. (11):

$$CO_3^= = K_{c2}' \cdot HCO_3^-/H^+ \tag{31}$$

Substituting Eq. (31) into Eq. (7):

Alkalinity = $HCO_3^- + K_{c2}' \cdot HCO_3^-/H^+ + OH^- - H^+$

and substituting Eq. (12) into this equation and solving for HCO_3^-:

$$HCO_3^- = \frac{\text{Alkalinity} - K_{cW}'/H^+ + H^+}{1 + K_{c2}'/H^+} \tag{32}$$

Equating Eq. (30 and 32), the general equilibrium equation in terms of Alkalinity, Acidity and H^+, is derived, i.e.

$$(1 + K'_{c2}/H^+)(Acidity - H^+ + K'_{cW}/H^+)$$

$$= (1 + H^+/K'_{c1}) \, (Alkalinity - K'_{cW}/H^+ + H^+) \tag{33}$$

For measured values of any two of the parameters Alkalinity, Acidity and H^+, the third parameter is uniquely defined.

To obtain a graphical plot of Eq. (33) with Alkalinity and Acidity as co-ordinate axes, the equation is rearranged into a more convenient form:

$$Acidity - H^+ + K_{cW}/H^+ = \left(\frac{1 + H^+/K'_{c1}}{1 + K'_{c2}/H^+}\right).(Alkalinity - K'_{cW}/H^+ + H^+)$$

Putting $Y = \dfrac{1 + H^+/K'_{c1}}{1 + K'_{c2}/H^+}$ in the above equation and rearranging:

$$Acidity = Y.Alkalinity + (Y + 1)(H^+ - K'_{cW}/H^+) \tag{34}$$

Equation (34) shows that for some chosen fixed value of pH (and hence some fixed value of H^+) there is a linear relationship between Alkalinity and Acidity (see Figure 1).

A complete program print-out is presented in Appendix D for plotting the pH-Alkalinity-Acidity conditioning diagram for any selected ionic strength and temperature. In Appendix B are a series of pH-Alkalinity-Acidity charts covering a range of combinations of ionic strength and temperature likely to be encountered in water treatment practice.

A plot of pH against Alkalinity and Acidity using Eq. (34) has four quadrants (see Figure 3). The significance which can be attached to a water whose condition plots in a particular quadrant is discussed below. In considering the diagram as a whole it is

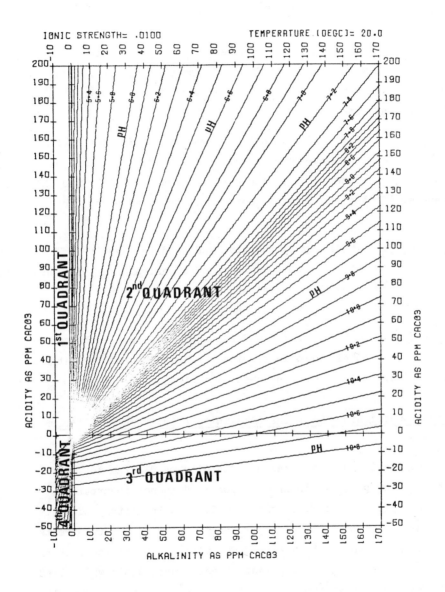

Figure 3. 4 Quadrants for the conditioning diagram

necessary to point out that it is assumed that only the carbonic system and strong acids and bases are present in the water. If another weak acid-base system is present, for example ammonia, the relationships depicted will not be correct.

Quadrant 1: Alkalinity is negative and Acidity positive. The pH of such a water will be below the carbonic acid equivalence point, i.e. the water contains a strong acid and pH change is no longer affected by the carbonic system.

Quadrant 2: Alkalinity and Acidity both positive. The pH of such a water lies somewhere between the carbonic acid and carbonate equivalence points. The carbonic system contols the pH changes. This quadrant will contain all the natural surface waters.

Quadrant 3: Alkalinity is positive and Acidity negative. The pH of such a water is above its carbonate equivalence point and indicates the presence of a strong base. The carbonic system has little effect on the pH of the system.

Quadrant 4: Both Alkalinity and Acidity negative. Such a water would have a total carbonic species concentration of less than 10^{-7} moles/ℓ and need not concern us further.

Water Conditioning Using the Alkalinity-Acidity-pH Equilibrium Diagram

The application of the Alkalinity-Acidity-pH diagram to water conditioning problems is limited for it does not incorporate the precipitation of $CaCO_3$ when the $CaCO_3$ solubility product is exceeded. In some problems, for example blending of two treated waters, it is very likely that $CaCO_3$ will not precipitate even though the final water be slightly oversaturated, in which event the diagram is most useful in determining the final blended water condition. With the main proviso - regarding $CaCO_3$

precipitation - the diagram can be useful for a number of purposes:

1. Predicting Alkalinity, Acidity and pH of a water after
    chemical dosing.

2. Blending of treated waters.

3. A general method for determining Alkalinity and Acidity.

1. <u>Chemical Dosing of Water where no $CaCO_3$ Precipitation Occurs</u>

    <u>Examples</u>:

(a) Water has the following quality: Alkalinity 60 ppm as $CaCO_3$,
    pH 9,4 and there is no Ca^{++} present; using the equilibrium
    diagram, Figure 4, predict Alkalinity, Acidity and pH of
    the water after sequentially adding:

    (i)   10 ppm NaOH (as $CaCO_3$)

    (ii) 10 ppm Na_2CO_3 (as $CaCO_3$)

    (iii) 30 ppm CO_2 (as $CaCO_3$)

        (i)   Initial state of the water plots at Point 1 in
              Figure 4. The vertical ordinate value of Point
              1 gives the initial Acidity in the water, i.e.
              Acidity 47 ppm as $CaCO_3$.

              Referring to Table 2 adding 10 ppm NaOH increases
              Alkalinity by 10 ppm and decreases Acidity by
              10 ppm (all concentrations in ppm as $CaCO_3$), i.e.

              New Alkalinity = (60 + 10) = 70 ppm as $CaCO_3$

              New Acidity   = (47 - 10) = 37 ppm as $CaCO_3$

              These two values plot at Point 2 in Figure 4,
              i.e. new pH value is 9,9.

        (ii) Referring to Table 2: adding 10 ppm Na_2CO_3
              increases Alkalinity by 10 ppm (both as ppm

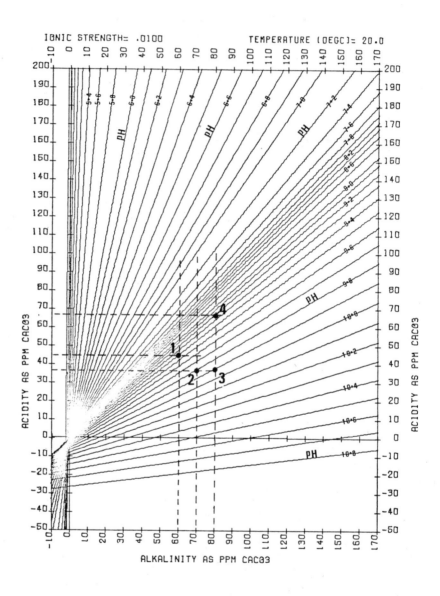

Figure 4. Sequential addition of 10 ppm NaOH, 10 ppm Na$_2$CO$_3$ and 30 ppm CO$_2$ (all in ppm CaCO$_3$) to a water (Point 1)

$CaCO_3$) and Acidity does not change, i.e.

New Alkalinity = (70 + 10) = 80 ppm as $CaCO_3$

New Acidity   = (37 + 0) = 37 ppm as $CaCO_3$

The state of this water plots at Point 3 in
Figure 4 with new pH 10,0.

(iii) Referring to Table 2: adding 30 ppm CO_2 to the
water increases the Acidity by 30 ppm (both as
$CaCO_3$) and Alkalinity does not change, i.e.

New Alkalinity = (80 + 0) = 80 ppm as $CaCO_3$

New Acidity   = (37 + 30) = 67 ppm as $CaCO_3$

The new state of the water plots at Point 4 in
Figure 4 with new pH value 9,3.

2. Water Blending

Three waters A, B and C, whose Alkalinity and pH values are given
below, are blended in the ratio: 4 parts water A to 3 parts water
B to 2 parts water C. Using an equilibrium diagram, Figure 5,
determine the condition of the blended water.

	pH	Alkalinity ppm as $CaCO_3$	Acidity (obtained from Figure 5) ppm as $CaCO_3$
Water A	7,0	30	46
Water B	9,0	60	55
Water C	9,4	80	64

Plotting the pH and Alkalinity values for each water in Figure
5 gives their respective Acidity values as listed in the table
above.

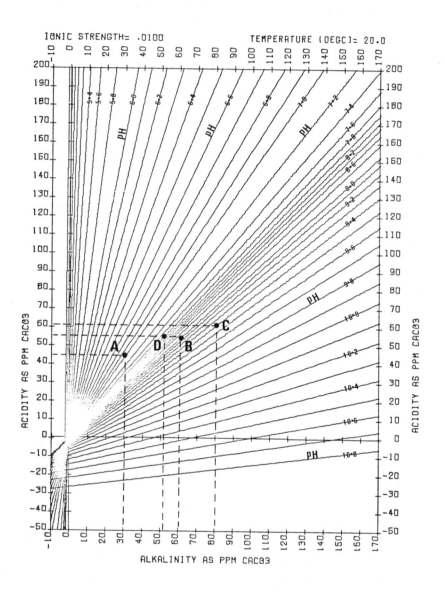

Figure 5. Water blending: 4 parts of water A, 3 parts B, and 2 parts C gives water D

The Alkalinity and Acidity values for the blended water are given by the sum of the fractional contributions of the three waters, i.e.

Alkalinity of blended water $= \frac{4}{9} \cdot 30 + \frac{3}{9} \cdot 60 + \frac{2}{9} \cdot 80$

$$= 51 \text{ ppm as } CaCO_3$$

Acidity of blended water $= \frac{4}{9} \cdot 46 + \frac{3}{9} \cdot 55 + \frac{2}{9} \cdot 64$

$$= 53 \text{ ppm as } CaCO_3$$

Water with Alkalinity 51 ppm and Acidity 53 ppm (both as $CaCO_3$) plots at Point D in Figure 5, i.e.

pH of blended water $= 8,2$

3. Measurement of Alkalinity and Acidity using the Alkalinity-Acidity-pH Equilibrium Diagram

The Alkalinity-Acidity-pH diagram, Figure 1, can be used to measure Alkalinity or Acidity of a water without predetermining the $H_2CO_3^*$ equivalence point. The method requires adding a known mass of base or acid to a water and accurately noting the pH change. This determination is simple and quick and can be used with equal facility both in the laboratory and in the field. The method is particularly useful for waters with low Alkalinity where large relative errors result from titrating to an incorrect endpoint. However, in the technique of measurement care should be taken that during the addition of the acid no appreciable CO_2 exchange occurs between the sample and the air. If a strong base is added the solubility product for $CaCO_3$ should not be exceeded.

These two effects are discussed in more detail once the method itself has been established. The procedure for the determination is briefly outlined as follows:

(a) The initial pH is noted, a selected mass of standard strong
     acid or base is added to a water sample and the final pH
     noted.

(b) Referring to Table 2: adding say 10 ppm (as $CaCO_3$) of strong
     acid to a water increases Acidity and decreases Alkalinity
     each by 10 ppm as $CaCO_3$. This change is described in the
     conditioning diagram by a vector $45°$ upwards to the left
     and of length $\sqrt{10^2 + 10^2} = \sqrt{200}$ units. (If a strong base is
     added its vector would be equal and opposite to that for the
     strong acid).

(c) The base of the vector described in (b) above is now moved
     along the initial pH line, the vector being shifted parallel
     to itself until the head of the vector touches the final pH
     line. These 'points of touch' on the initial and final pH
     lines represent the initial and final states of the water.
     In particular, the initial state is defined, i.e. Alkalinity,
     Acidity and pH. Knowing the Alkalinity and Acidity the
     total carbonic species concentration is determined from Eq.
     (8), Chapter 4.

   The method was tested experimentally on a number of water
samples with different pH and Alkalinity values. For each sample
the Alkalinity was determined from an automatic plot of the pH-
acid added curve. The $H_2CO_3^*$ equivalence point was determined
from the curvature of this graphical plot. From the pH-acid
added plot the initial and final pH for a selected mass of acid
addition was read off and the Alkalinity determined from the
Alkalinity-Acidity-pH equilibrium diagram. Results are compared
in Table 3.
   The method and solution procedure are illustrated in the
following example.

Table 3.

Initial pH	Final pH	Acid added ppm as $CaCO_3$	Alkalinity (graphical) ppm as $CaCO_3$	Alkalinity (by titration) ppm as $CaCO_3$
8,07	7,07	15	100	100
7,62	7,12	5	53	50
10,40	9,90	20	75	75
10,30	9,80	19	76	75
7,35	6,85	3,5	21	20

Water temperature 25°C; $\mu \approx 0,01$

The ionic strength of a water is 0,001, temperature 15°C. The initial pH is 7,0. After adding exactly 5 ppm HCl (as $CaCO_3$) the pH of the water is 5,6. Using the equilibrium diagram calculate the initial Alkalinity and Acidity of the water.

(a) pH lines 7,0 and 5,6 are shown drawn accentuated in the equilibrium diagram, Figure 6.

(b) Vector AB, equal to $\sqrt{25} + 25 = \sqrt{50}$, is drawn in the diagram at 45° and moved parallel to itself, with its base on the pH 7,0 line, until it eventually intersects pH 5,6 line at Point D.

(c) Point D defines the state of the water after adding 5 ppm (as $CaCO_3$) of strong acid. Thus the vector CD describes the change in the condition of the water due to acid addition, and Point C defines the initial condition of the water which is therefore: Alkalinity 6,0 ppm, Acidity 9,0 and pH 7,0.

In applying this graphical technique the user must be vigilant to two sources of error:

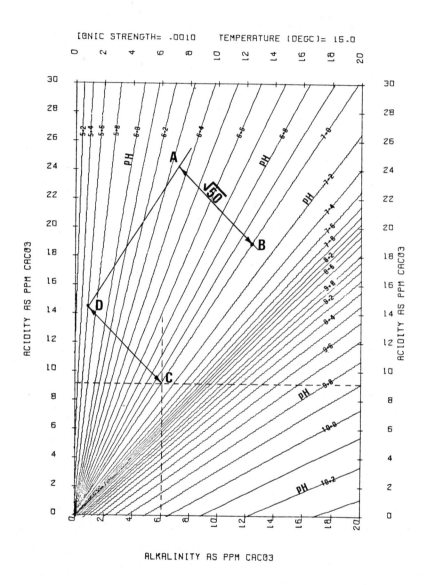

Figure 6.   Initial pH is 7,0; addition of 5 ppm HCl alters pH to
5,6 giving initial Alkalinity as 6 ppm

(i) CO_2 exchange during the titration. If an appreciable con-
      centration of CO_2 is exchanged between the sample and air
      after noting the initial pH (i.e. after adding the acid or
      base) the results will be in error. The reason is that CO_2
      transfer causes a change in the Acidity added over and above
      that due to the acid or base added. Thus the vector descri-
      bing the acid or base added is no longer a known parameter
      and merely using that vector describing the acid added in
      the diagram gives an incorrect estimate of Alkalinity,
      Acidity and C_T.

      The effects of carbon dioxide exchange can be limited to an
      insignificant level if the following points are observed:

(a)   The sample should be stirred smoothly and not too
      vigorously.

(b)   The initial pH value should be noted just before
      the strong acid or base is added. Should the CO_2
      be lost or gained by the sample prior to noting
      the initial pH, the Alkalinity value given by this
      method is not affected. The Acidity and C_T values
      obtained from the diagram will be those values
      after CO_2 stripping prior to the test. Thus if
      the Acidity and C_T values for the water prior to
      CO_2 stripping are required - they are given by
      the point of intersection in the diagram of the
      measured Alkalinity value and the pH in the water
      prior to CO_2 stripping.

      Thus for a water heavily oversaturated with CO_2
      (say a groundwater) it is advisable first to
      agitate the water until the pH reaches a stable
      value before noting the initial pH and adding the
      standard acid or base.

(ii)   If a standard strong base is used to bring about the pH
       change in this method of Alkalinity measurement, there
       is a possibility that the $CaCO_3$ solubility product be
       exceeded causing formation of $CaCO_3$ nucleii and erro-
       neous results. The solubility of $CaCO_3$ in *pure* water
       is only about 20 ppm (i.e. 20 ppm as $CaCO_3$ of each Ca^{++}
       and Alkalinity) so that for most natural waters addition
       of a strong base, say NaOH, will cause $CaCO_3$ oversaturation
       if the pH is raised appreciably. Thus estimation of Alka-
       linity by addition of a strong base should only be used
       with very soft natural waters, with Alkalinity and Ca^{++}
       each less than about 15 ppm as $CaCO_3$. The concentration
       of a strong base added should not exceed about 5 ppm as
       $CaCO_3$ or the pH raised above about 7.

In water conditioning one is not only concerned with adjusting
pH and Alkalinity to some desired values but also the calcium
and/or magnesium contents must not exceed some limiting concentration, i.e. the water must not be too hard. Furthermore, the
water must be saturated or slightly oversaturated with respect
to $CaCO_3$.

Calculations for estimating chemical dosages to adjust the
water to meet these requirements must consider equilibrium between two phases: the Ca^{++} and carbonic species in the aqueous
phase (defined by Ca^{++}, Alkalinity and pH) and $CaCO_3$ in the solid
phase.

In Chapter 4 it was shown that for a water just saturated with
$CaCO_3$ the calcium concentration can be calculated from the pH
and Alkalinity values in the water, Eq. (16), Chapter 4. Thus in
general, for any particular state of equilibrium between aqueous
species defined by any two of the parameters C_T, Alkalinity,
Acidity and pH, there is some fixed corresponding calcium value
for $CaCO_3$ saturation. Superimposing the saturated calcium value

for each state of equilibrium in the Acidity-Alkalinity-pH diagram gives a two phase equilibrium diagram (see Figures 7 and 8).

The calcium values for $CaCO_3$ saturation can be plotted in a diagram with co-ordinate axes Alkalinity and Acidity as follows:

$$\text{Alkalinity} = \frac{H^+ \cdot K'_{cS}}{K'_{cS} \cdot Ca^{++}} + \frac{K'_{cS}}{Ca^{++}} + \frac{K'_{cW}}{H^+} - H^+ \tag{35}$$

and substituting Eqs. (10 to 13) into Eq. (8):

$$\text{Acidity} = \frac{H^{+2} \cdot K'_{cS}}{Ca^{++} \cdot K_{c1} \cdot K'_{c2}} + \frac{H^+ \cdot K'_{cS}}{K'_{c2} \cdot Ca^{++}} + H^+ - \frac{K'_{cW}}{H^+} \tag{36}$$

Solving for H^+ in Eq. (35):

$$H^+ = \frac{\text{Alk} - K'_{cS}/Ca^{++} + \sqrt{(\text{Alk} - K'_{cS}/Ca^{++})^2 - 4(K'_{cS}/K'_{c2} \cdot Ca^{++} - 1) \cdot K'_{cW}}}{2K'_{cS}/(K'_{c2} \cdot Ca^{++})} \tag{37}$$

and substituting Eq. (37) into Eq. (36) gives a general two phase equilibrium equation for $CaCO_3$ saturated water in terms of the parameters Alkalinity, Acidity and Ca^{++} concentration. To plot a saturated calcium line for Ca^{++} at some specified value, determine H^+ for a series of Alkalinity values from Eq. (37) and substitute in Eq. (36) to find the corresponding Acidity values. By plotting the calculated values of Acidity against Alkalinity, a line is traced out of the specified Ca^{++} value (see Figure 7).

If a water is just saturated with respect to $CaCO_3$ the lines representing the measured calcium, pH and Alkalinity values will intersect at a point in the diagram. However, if the calcium line does not pass through the point describing equilibrium between aqueous species (i.e. Alkalinity, pH intersection) in the diagram, the water is either over- or undersaturated with respect to $CaCO_3$. The saturation state of the water is deduced by noting the value of the calcium line passing through the aqueous phase

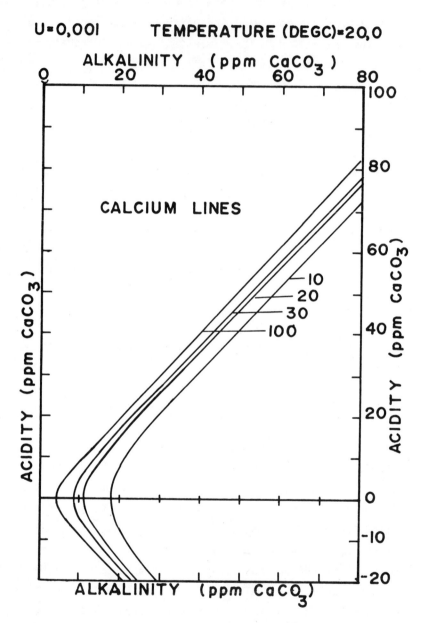

Figure 7. Family of lines representing Ca^{++} values for CaCO$_3$
saturation.

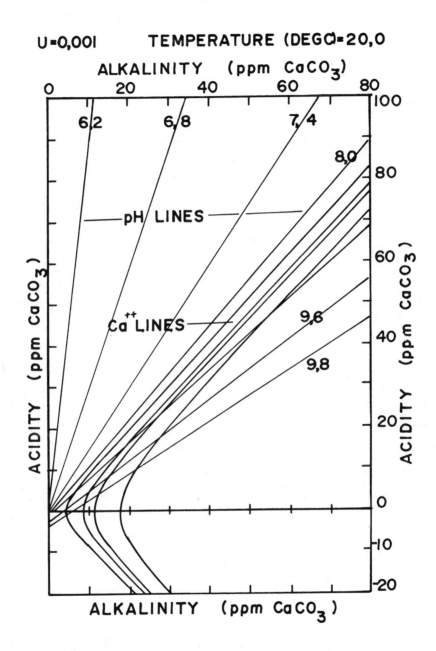

Figure 8. Ca^{++} and pH lines for a two phase conditioning diagram

equilibrium point and comparing this theoretical calcium value
with the measured calcium value. If the *theoretical* Ca value is
greater than the *measured* value the water is *undersaturated* and
vice versa.

The Alkalinity-Acidity-pH-Ca diagram for two phase equilibrium
is unsuitable to use for most practical problems. From Figure 8
in the region of the diagram where Alkalinity and Acidity are
both positive, not only do the pH and Ca lines have the same sign
of slope so that intersection is ill defined, but also in the pH
region 8 to 9 the pH and Ca lines plot indistinguishably close
together.

However, an understanding of the concepts underlying the development of this diagram, and the graphical approach in solving
two phase conditioning problems in this diagram, gives insight
into the fundamentals upon which the very useful Modified Caldwell-Lawrence conditioning diagram is based.

Graphical solutions to two types of problems frequently encountered in practice are outlined below. Solution to these
problems illustrates that not only is the Alkalinity-Acidity-pH-
Ca equilibrium diagram impractical because of the lack of clarity
between the pH and Ca lines but also solution to problems can
only be obtained by methods of trial and error.

1. To ascertain graphically the $CaCO_3$ saturation state of a
    water: For a measured Alkalinity, pH and Ca value of a water
    the two phase diagram can be used directly to give the Lange-
    lier Saturation Index of the water, and, by a method of trial
    and error, to give a quantitative measure of the mass of
    $CaCO_3$ which will either precipitate or dissolve into the
    water to achieve $CaCO_3$ saturation.

2. Water conditioning problems: A concentration of some chemical
    (say NaOH) is added to water with a measured Alkalinity, pH
    and calcium concentration. It is required to predict the

final state which the water achieves with time. For problems
of this kind the graphical solution procedure is usually
carried out in two steps:

(a)   the new pH, Alkalinity and Ca values are calculated
      by regarding $CaCO_3$ as infinitely soluble, i.e. on
      the basis of a single phase model.

(b)   the $CaCO_3$ saturation state of the water is ascertained
      and if oversaturated, the final saturated state which
      the water achieves with time is calculated by a method
      of trial and error as in (1) above.

Thus, in effect, the method of graphical solution to two phase
problems considers $CaCO_3$ precipitation really only as a consequence of changes that have already occurred in pH and Alkalinity
(or Acidity).

Example 1

Analysis of a water sample gives: Alkalinity 30 ppm, calcium 45
ppm (both as $CaCO_3$) and pH 9,2. Using a two phase equilibrium
conditioning diagram determine; (a) the Langelier Saturation Index of the water; (b) the mass of $CaCO_3$ which either precipitates
or dissolves into the water with time.

(a)   Initial pH and Alkalinity plot as Point 1 in Figure 9 (i).
      The value of the calcium line passing through Point 1 gives
      the theoretical Ca value required for saturation, i.e. 25
      ppm as $CaCO_3$. This value is less than the measured calcium
      value (45 ppm as $CaCO_3$) and one deduces that the water is
      oversaturated and $CaCO_3$ precipitation occurs with time.

      Langelier's Saturation Index, S.I. is defined in Chapter 4
      as: $S.I. = pH - pH_s$.

      pH_s is defined as a theoretical pH value which gives satura-
      tion for the measured Alkalinity and calcium concentrations

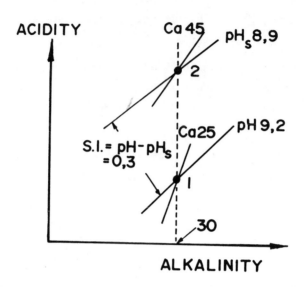

Figure 9(i). Determination of the Langelier S.I. in Alkalinity-
Acidity-pH-Ca diagram

in the water, i.e. pH_s is the value of the pH line passing
through the intersection of the measured values of calcium
and Alkalinity in Figure 9(i) i.e. pH_s is the value of the
pH line through Point 2 and has the value pH_s = 8,9.

Thus S.I. = 9,2 - 8,9 = +0,3 and by definition a positive
index indicates an oversaturated water.

(b) To estimate the mass of $CaCO_3$ which will ultimately precipitate from this water one must first predict the final saturated state which the water achieves with time. This is
achieved as follows:

When $CaCO_3$ precipitates from an oversaturated water the
following two conditions must be satisfied:

(i) From Table 2: loss of $CO_3^=$ ions from solution has no effect on the Acidity, i.e. the Acidity of the oversaturated water and its final condition at saturation are equal.

(ii) When $CaCO_3$ precipitates from an oversaturated water equal changes occur in the calcium and Alkalinity values. This is illustrated as follows:

Each equivalent of $CaCO_3$ precipitating removes one equivalent each of Ca^{++} and $CO_3^=$ from solution, i.e.

$$Ca^{++} \text{ precipitated} = CO_3^= \text{ precipitated} \qquad (38)$$

and from Table 2

$$CO_3^= \text{ precipitated} = \text{Alkalinity precipitated} \qquad (39)$$

and equating Eqs. (35 and 36)

$$Ca^{++} \text{ precipitated} = \text{Alkalinity precipitated} \qquad (40)$$

Interpreting these two requirements in a conditioning diagram (Figure 9(ii)) for the problem under discussion:

(i) The final saturated state of the water lies on the horizontal Acidity line through Point 1 (i.e. Acidity 26 ppm as $CaCO_3$).

(ii) Using a method of trial and error, the second criterion (ii) above is used to fix the final saturated state on the Acidity line through Point 1 as follows:

Assume 2 ppm $CaCO_3$ has precipitated from the water. The Alkalinity is now $(30 - 2) = 28$ ppm as $CaCO_3$. The new equilibrium state of the water is therefore given by the intersection of the Acidity 26 ppm line and the Alkalinity 28 ppm line, i.e. Point 3, Figure 9(ii). The calcium concentration required for saturation at this new equilibrium point is given by the value of the

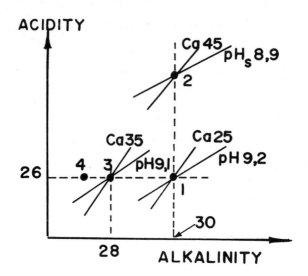

Figure 9(ii). Estimating the mass of $CaCO_3$ which ultimately precipitates from a supersaturated water

     calcium line passing through Point 3, i.e. Ca 35 ppm as $CaCO_3$. Now, noting that after 2 ppm $CaCO_3$ has precipitated the calcium value is (45 - 2) = 43 ppm as $CaCO_3$, one deduces that the water is still over-saturated.

     Proceeding in this manner of trial and error Point 4 in Figure 9 is established as the final saturated state of the water, i.e. Alkalinity 26 ppm, calcium 41 ppm (both as $CaCO_3$) and pH 9,0.

     Thus the mass of $CaCO_3$ which will ultimately precipitate from the water is given by the difference between either the initial and final Alkalinity or calcium values, i.e. 4 ppm as $CaCO_3$.

Example 2

Analysis of a water sample gives: Alkalinity 25 ppm, calcium 20 ppm (both as $CaCO_3$) and pH 9,0. Using an equilibrium condition-ing determine the state of the water after 10 ppm (as $CaCO_3$) of $Ca(OH)_2$ is added.

   Conditioning problems of this kind are solved in two steps:

(a)   Consider $CaCO_3$ as being infinitely soluble in water and
      on this basis graphically ascertain the new state of
      equilibrium between species in the squeous phase.

      Alkalinity 25 ppm as $CaCO_3$ and pH 9,0 plots at Point 1 in
      Figure 10, i.e. initial Acidity in the water is 24 ppm as
      $CaCO_3$.

      Referring to Table 2, adding 10 ppm $Ca(OH)_2$ as $CaCO_3$ to
      the water has the following effect:

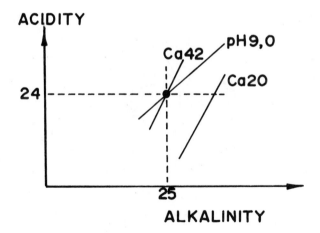

Figure 10.   Plotting the condition Alkalinity 25 ppm, Ca 20 ppm
(as $CaCO_3$) and pH 9,0 indicates undersaturation

Initial Alkalinity = 25 ppm as $CaCO_3$
Alkalinity added = 10 ppm as $CaCO_3$
i.e. Final Alkalinity = 35 ppm as $CaCO_3$

Initial Acidity = 24 ppm as $CaCO_3$
Acidity added = -10 ppm as $CaCO_3$
Final Acidity   = (24 - 10) = 14 ppm as $CaCO_3$

Initial Ca = 20 ppm as $CaCO_3$
Ca added = 10 ppm as $CaCO_3$
Final Ca   = (20 + 10) = 30 ppm as $CaCO_3$

The new equilibrium state of the water is defined by the
final values of Alkalinity and Acidity given above. Plotting those values in Figure 11 gives the final pH value
as 10,1, i.e. Point 2. The calcium concentration re
quired for saturation is now given by the value of the
calcium line passing through Point 2, i.e. Ca 5 ppm as

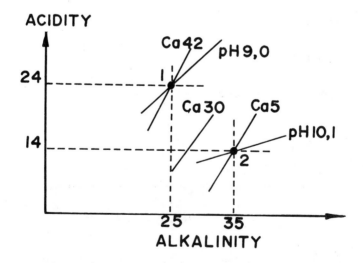

Figure 11. 10 ppm $Ca(OH)_2$ is added to a water plotting at Point 1

$CaCO_3$. However, the actual calcium value is 30 ppm as $CaCO_3$, i.e. the water is oversaturated and $CaCO_3$ precipitates.

(b) Using a similar method of trial and error as in Example 1 above, the final saturated state of the water is found graphically as: Alkalinity 20 ppm, Ca 15 ppm (both as $CaCO_3$) and pH 9,75.

Modified Two Phase Equilibrium Diagram

The examples above illustrate that solution to $CaCO_3$ precipitation problems in the Alkalinity-Acidity-pH-Ca equilibrium diagram are only possible using methods of trial and error. A direct graphical solution to precipitation problems is possible by introducing into the equilibrium diagram an extra parameter describing a necessary condition (over and above that of constant Acidity) to be satisfied for $CaCO_3$ precipitation. This extra parameter is chosen as follows:

For $CaCO_3$ precipitation from an oversaturated water, from Eq. (37),

$$Alkalinity_{precipitated} = Ca_{precipitated}$$

i.e. Δ Alk = Δ Ca

i.e. Δ Alk - Δ Ca = 0

i.e. Alk - Ca = constant

$$= C_2 \tag{41}$$

From Eq. (41) for $CaCO_3$ precipitation from an oversaturated water the difference between the Alkalinity and calcium concentrations remains constant.

From solubility and aqueous equilibria relationships one can develop an equation for C_2 (i.e. the parameter Alk - Ca) in terms

of the parameters Alkalinity and Acidity and superimpose this
parameter on the Alkalinity-Acidity-pH-Ca equilibrium diagram.
This new equilibrium diagram involving the parameters Alkalinity,
Acidity, pH, Ca and (Alk - Ca) can be used to solve directly any
single or two phase problem for the calcium-carbonic system.

An equation to be used for superimposing the parameter (Alk-
Ca) on the equilibrium diagram with Alkalinity and Acidity as
co-ordinates is developed as follows:

Substituting Eq. (13) into Eq. (41)

$$Alk - K'_{cS}/CO_3^= = C_2$$

and rearranging this equation

$$CO_3^= = K'_{cS}/(Alk - C_2) \qquad (42)$$

and from Eqs. (15 and 17) Chapter 3, with concentrations expressed in ppm as $CaCO_3$:

$$CO_3^= = \frac{K'_{c2}}{H} \cdot \frac{C_T}{1 + H^+/2K'_{c1} + K'_{c2}/2H^+} \qquad (43)$$

Equating Eqs. (42 and 43) and solving for C_T

$$C_T = \frac{K'_{cS}}{(Alk - C_2)} \cdot (1 + H^+/2K'_{c1} + K'_{c2}/2H^+) \cdot H^+/K'_{c2} \qquad (44)$$

and rewriting Eq. (53) Chapter 3, with species concentrations
expressed in ppm as $CaCO_3$:

$$C_T = \frac{(Alk - OH^- + H^+)}{1 + K'_{c2}/H^+} \cdot (1 + K'_{c2}/2H^+ + H^+/2K'_{c1}) \qquad (45)$$

Equating Eqs. (44 and 45):

$$Alk - K'_{cW}/H^+ + H^+ = \frac{K'_{cS} \cdot H^+}{(Alk - C_2) \cdot K'_{c2}} \qquad (46)$$

Writing $K'_{cS}/(Alk - C_2) = X$ in the above equation and solving for H^+:

$$H^+ = \frac{(X - Alk) \pm \sqrt{(X - Alk)^2 - 4(X/K'_{c2} - 1) K'_{cW}}}{2(1 - X/K'_{c2})} \qquad (47)$$

For any fixed chosen value of C_2, the parameters Alkalinity and Acidity are directly related through Eqs. (34 and 47) and this relationship can be plotted in an equilibrium diagram with Alkalinity and Acidity as Cartesian co-ordinates (see Figure 12 (i)).

Water conditioning problems in which $CaCO_3$ precipitation occurs are now solved directly in the modified equilibrium diagram, (Figure 12(ii)). The particular problems of predicting both the mass of $CaCO_3$ which ultimately precipitates from an oversaturated water and the final saturated state for such a water are solved in the diagram considering the following requirements:

(a) Acidity in a water remains unchanged with $CaCO_3$ precipitation. (This is appreciated by referring to Table 2).

(b) The difference between the Alkalinity and calcium concentrations (i.e. the parameter C_2) remains unchanged with $CaCO_3$ precipitation).

(c) For a saturated water the lines for the values of all the parameters Alkalinity, Acidity, pH, Ca and (Alk-Ca) intersect in a point in the diagram. Thus at saturation, intersection of the lines for values of any two of those parameters (in particular Acidity and Alk-Ca) defines the saturated state of the water in the equilibrium diagram.

Thus, from (a) to (c) above, the final saturated state which an oversaturated water ultimately achieves is given by the intersection of the lines representing the 'oversaturated' values of Acidity and (Alk-Ca) in the diagram.

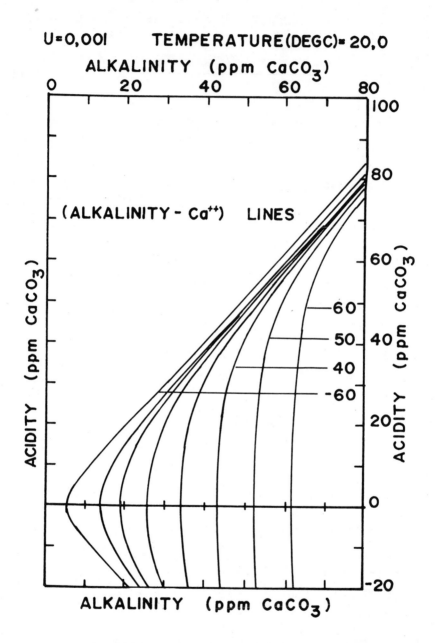

Figure 12(i). Lines representing (Alk-Ca) at $CaCO_3$ saturation in a diagram with co-ordinates Alkalinity and Acidity.

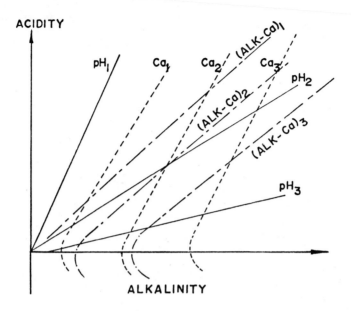

Figure 12(ii). Lines representing pH, Ca and (Alk-Ca) at $CaCO_3$
saturation

Example

Analysis of a water gives: Alkalinity 35 ppm, calcium 60 ppm (both
as $CaCO_3$) and pH 9,0. What final saturated state will the water
achieve with time?

Alkalinity and pH of the water plot at Point 1 in Figure 13.
The value of the calcium line passing through Point 1 is 30 ppm
as $CaCO_3$, i.e. the water is oversaturated with respect to $CaCO_3$.

Initial Acidity of the water (i.e. Acidity value of Point 1
in the diagram) is 29 ppm as $CaCO_3$. Interpreting the requirements, (a), (b) and (c) above, for $CaCO_3$ precipitation in the
equilibrium diagram (Figure 13):

(a) The final state of the water has an Acidity of 29 ppm
     as $CaCO_3$, i.e. lies on the horizontal line through
     Point 1.

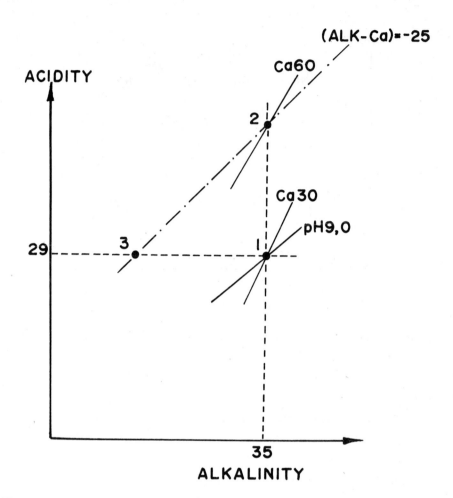

Figure 13.   Precipitation of $CaCO_3$ from a supersaturated water.
Point 3 defines the final saturated state

(b)   The initial (Alk-Ca) value for the water is (35 - 60) =
      -25 ppm as $CaCO_3$.   Thus the final state of the water
      lies on the (Alk-Ca) line equal to -25 ppm.

(c)   Intersection of the Acidity 29 ppm line and (Alk-Ca)
      line of -25 ppm defines the final saturated state of
      the water in the diagram, i.e. Point 3 with Alkalinity

33 ppm, calcium 57 ppm (both as $CaCO_3$) and pH 8,8.

The two phase equilibrium diagram with Alkalinity and Acidity as co-ordinate axes has limited practical use because of the difficulty in distinguishing between pH and Ca lines in the pH range 8 to 9. Plotting lines for (Alk-Ca) in the diagram worsens the situation. The value of this new equilibrium diagram is to illustrate that by graphically inter-relating the parameters Alkalinity, Acidity, pH, Ca and (Alk-Ca) using equilibria relationships, direct solution to two phase problems is possible. In the following section these 5 parameters are again considered in a graphical plot. Choosing Acidity and (Alk-Ca) as the co-ordinate axes and plotting the parameters Alkalinity, pH and Ca in the diagram gives the modified Caldwell-Lawrence conditioning diagram which is very useful in its practical application to two phase problems.

MODIFIED CALDWELL-LAWRENCE DIAGRAM

The Modified Caldwell-Lawrence diagram graphically presents equilibria inter-relationships between Alkalinity, Acidity, pH and calcium in a plot with the parameters Acidity and (Alk-Ca) as co-ordinates. The diagram is used for chemical conditioning of water considering only equilibrium in the aqueous phase, or aqueous and solid phases, or aqueous and gas phase, or aqueous, solid and gaseous phases. Practical application includes water softening, water stabilization, water blending and CO_2 stripping problems.

This diagram incorporates the same five parameters as the two phase equilibrium diagram discussed in the previous section, i.e. Acidity, C_2, Alkalinity, pH, Ca^{++}. The method of solution is also essentially the same. However, in the Modified Caldwell-Lawrence diagram the advantages in choosing the parameters Acidity and (Alk-Ca) as co-ordinates instead of Acidity and Alkalinity is that the lines representing the Alkalinity, calcium and pH give good intersection (see Figures 14a and 14b). This was not the case for the two phase equilibrium diagram using Alkalinity and Acidity as co-ordinates (see Figure 12).

The co-ordinate system selected for the Modified Caldwell-Lawrence diagram is shown in Figure 15., i.e. (Alk-Ca) increases in the positive X direction and Acidity increases in the negative Y direction.

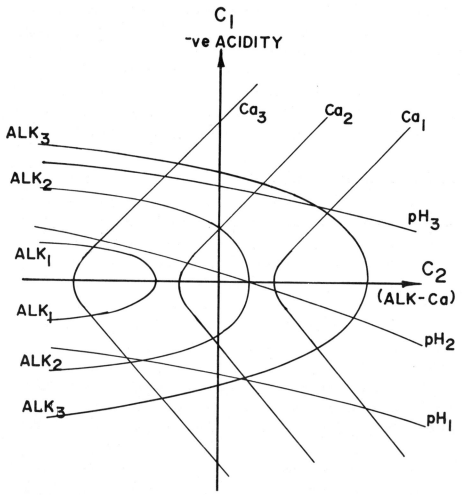

Figure 14a. Modified Caldwell-Lawrence Conditioning Diagram

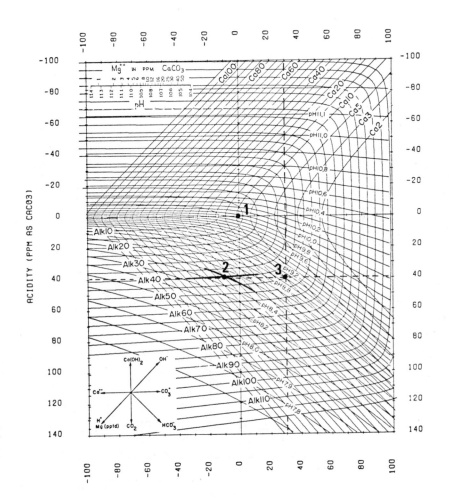

IONIC STRENGTH= .0010   TEMPERATURE (DEGC)= 25.0

APPROXIMATE TDS(PPM)= 40

Figure 14b.   Origin of diagram defines solubility of CaCO$_3$ in
water

Define:

$C_1 = -$ Acidity   i.e.

$C_1 = OH^- - H^+ - HCO_3^- - HCO_3^*$ and   (48)

$C_2 = (Alk-Ca)$ (49)

By definition C_1 is negative Acidity. The parameters C_1 and C_2 increase in the positive Y and X directions respectively.

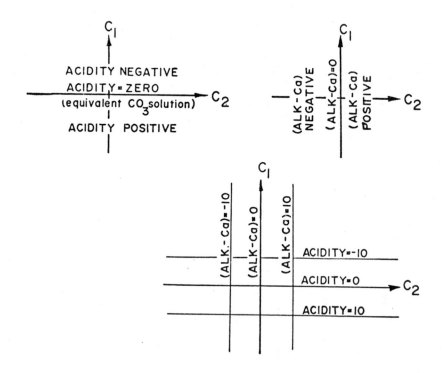

Figure 15. Co-ordinate system for Modified Caldwell-Lawrence Diagram

To represent saturated calcium equilibrium values for each of
the parameters Alkalinity, calcium and pH the solubility and
ionic equilibria relationships are used to derive equilibria
equations for each of the parameters pH, calcium and Alkalinity
in terms of the co-ordinate parameters C_1 and C_2. For explana-
tory purposes these equilibria relationships are depicted as
follows:

$$pH = f_1 (C_1, C_2)$$

$$Alk = f_2 (C_1, C_2)$$

$$Ca = f_3 (C_1, C_2)$$

Lines representing saturated equilibrium values for these parameters are now plotted in the diagram as follows:

Take for example the expression $Ca = f_3 (C_1, C_2)$. For a fixed
chosen value of Ca one has a direct relationship between C_1 and
C_2. Choosing a series of values for C_1 the corresponding values
of C_2 can be calculated. Plotting these corresponding co-ordinate
values gives a line representing the chosen Ca value for $CaCO_3$
saturated equilibrium conditions. Repeating this procedure for a
number of Ca values, and similarly for a number of Alkalinity and
pH values, gives the Modified Caldwell-Lawrence diagram (see
Figures 14a and 14b).

This diagram constitutes either a single or two phase equilibrium diagram. For a single aqueous system, any two of the parameters Alkalinity, Acidity and pH defines equilibrium values for
the carbonic species in solution. Thus in the Modified Caldwell-
Lawrence diagram the intersection point of the lines representing
the values of any two of these parameters defines the equilibrium
value of the third parameter, i.e. plotting the lines representing measured Alkalinity and pH values of a water in the diagram
the Acidity in the water is given by the Acidity ordinate value

of the intersection point. The value of the calcium line passing through this intersection point defines the required calcium concentration for $CaCO_3$ saturation for the particular pH and Alkalinity (or Acidity) values.

Equations for Plotting the Parameters Alkalinity, pH and Calcium in the Modified Caldwell-Lawrence Diagram

Equilibria equations relating each of the parameters pH, Alkalinity and calcium to the co-ordinate parameters C_1 and C_2 are derived from basic equilibria equations as follows:

1. Equations for pH Lines

Substituting Eq. (7) into Eq. (49):

$$C_2 = HCO_3^- + CO_3^= + OH^- - H^+ - Ca^{++} \tag{50}$$

From Eq. (13):

$$Ca^{++} = K'_{cS}/CO_3^= \tag{51}$$

Substituting Eqs. (11, 12 and 51) into Eq. (50):

$$C_2 = HCO_3^- + \frac{K'_{c2} \cdot HCO_3^-}{H^+} + \frac{K'_{cW}}{H^+} - H^+ - \frac{K'_{c2} \cdot H^+}{K'_{c2} \cdot HCO_3^-}$$

and rearranging this equation:

$$HCO_3^2(-1 - K'_{c2}/H^+) + HCO_3(C_2 - K'_{cW}/H^+ + H^+) + K'_{cS} \cdot H^+/K'_{c2} = 0$$

Solving for HCO_3^- in the above equation gives:

$$HCO_3^- = \frac{-(C_2 - K'_{cW}/H^+ + H^+) \pm \sqrt{(C_2 - K'_{cW}/H^+ + H^+)^2 - 4(-1 - K'_{c2}/H^+)(K'_{cS} \cdot H^+/K'_{c2})}}{2(-1 - K'_{c2}/H^+)} \tag{52}$$

Substituting Eqs. (10 and 12) into Eq. (48):

$$C_1 = K'_{cW}/H^+ - HCO_3^- - H^+ \cdot HCO_3^-/K'_{c1} - H^+ \tag{53}$$

Substituting Eq. (52) into Eq. (53) gives an equilibrium equation inter-relating the parameters H^+ (and hence pH) with C_1 and C_2. Choosing a number of values for C_2 and a fixed selected value of H^+ (and hence pH) the corresponding C_1 values are calculated. Plotting these corresponding values of C_1 and C_2 in Figure 14 traces a line for the selected pH value in the equilibrium diagram.

2. Equations for Ca^{++} Lines

From Eq. (13):

$$CO_3^= = K'_{cS}/Ca^{++} \tag{54}$$

Substituting Eq. (54) into Eq. (11):

$$H^+ = K'_{c2} \cdot Ca^{++} \cdot HCO_3^-/K'_{cS} \tag{55}$$

Substituting Eq. (55) into Eq. (48):

$$C_1 = \frac{K'_{cW} \cdot K'_{cS}}{HCO_3^- \cdot K'_{c2} \cdot Ca^{++}} - \frac{K'_{c2} \cdot HCO_3^{-2} \cdot Ca^{++}}{K'_{cS} \cdot K'_{cl}} - HCO_3^- - \frac{K'_{c2} \cdot Ca^{++} \cdot HCO_3^-}{K'_{cS}}$$

and rearranging this equation:

$$\frac{K'_{c2} \cdot Ca^{++} \cdot HCO_3^{-3}}{K'_{cS} \cdot K'_{cl}} + \left(1 + \frac{K'_{c2} \cdot Ca^{++}}{K'_{cS}}\right) \cdot HCO_3^{-2} + C_1 \cdot HCO_3^- - \frac{K'_{cW} \cdot K'_{cS}}{K'_{c2} \cdot Ca^{++}} = 0 \tag{56}$$

Substituting Eqs. (7 and 54) into Eq. (49):

$$C_2 = HCO_3^- + K'_{cS}/Ca^{++} + K'_{cW}/H^+ - H^+ - Ca^{++} \tag{57}$$

and substituting Eq. (55) into Eq. (57):

$$C_2 = HCO_3^- + K'_{cS}/Ca^{++} + \frac{K'_{cW} \cdot K'_{cS}}{HCO_3^- \cdot K'_{c2} \cdot Ca^{++}} - Ca^{++} - HCO_3^- \cdot K'_{c2} \cdot Ca^{++} \tag{58}$$

Calcium lines can now be plotted in the equilibrium diagram, Figure 14, as follows:

For a chosen fixed value of Ca^{++} and a range of values for C_1: the cubic equation, Eq. (56) in HCO_3^- has only one real root which can be calculated by an iterative method of trial and error. Substituting this HCO_3^- value into Eq. (58) gives the C_2 value corresponding to each C_1 value. Plotting the corresponding C_1 and C_2 values traces a line for Ca in the equilibrium diagram.

3. Equations for Alkalinity Lines

Alkalinity lines plot in the Modified Caldwell-Lawrence diagram as curves each with two horizontal limbs joined by a vertical section of curve (see Figure 14). To plot the horizontal section of curve accurately, for some chosen Alkalinity value C_1 values are calculated for corresponding C_2 values. For the vertical section of curve C_2 values are calculated for corresponding C_1 values.

(a) Equation for Horizontal Section of Alkalinity Curve

From Eq. (13):

$$Ca^{++} = K'_{cS}/CO_3^=$$

and substituting this equation into Eq. (49):

$$C_2 = Alk - K'_{cS}/CO_3^= \tag{59}$$

Substituting Eq. (11) into Eq. (59):

$$C_2 = Alk - \frac{K'_{cS} \cdot H^+}{K'_{c2} \cdot HCO_3^-}$$

and solving for HCO_3^-

$$HCO_3^- = \frac{K'_{cS} \cdot H^+}{K'_{c2}(Alk - C_2)} \tag{60}$$

Substituting Eq. (11) into Eq. (7):

$$Alk = HCO_3^- + K'_{c2} \cdot HCO_3^-/H^+ + K'_{cW}/H^+ - H^+ \tag{61}$$

and rearranging this equation:

$$H^2 \left\{ \frac{K'_{cS}}{K'_{c2}(Alk - C_2)} - 1 \right\} + H\left\{ \frac{K'_{cS}}{(Alk - C_2)} - Alk \right\} + K'_{cW} = 0$$

i.e.

$$H^+ = \frac{\{-X + Alk\} \overset{+}{-} \sqrt{(Alk - X)^2 - 4X.K'_{cW}/K'_{c2}}}{2(X/K'_{c2} - 1)} \tag{62}$$

where $X = K'_{cS}/(Alk - C_2)$

Substituting Eqs. (60 and 11) into Eq. (48):

$$C_1 = K'_{cW}/H^+ - H^2.X/K'_{c1}.K'_{c2} - H.X/K'_{c2} - H \tag{63}$$

Substituting Eq. (62) into Eq. (63) gives an equilibrium equation in terms of C_1, C_2 and Alkalinity.

For a chosen Alkalinity value and a series of C_2 values, the corresponding C_1 values are calculated. Plotting the corresponding C_1 and C_2 values gives a trace of the horizontal sections of the Alkalinity line in the diagram.

(b) Equation for the Vertical Section of the Alkalinity Curve

In the pH region greater than 4,5 Eq. (61) approximates to:

$$Alk = HCO_3^- + K'_{c2} \cdot HCO_3^-/H^+ + K'_{cW}/H^+$$

Rearranging this equation and solving for H^+:

$$H^+ = \frac{(K'_{c2} \cdot HCO_3^-) + K'_{cW}}{(Alk - HCO_3^-)} \tag{64}$$

Substituting Eq. (64) into Eq. (45):

$$C_1 = \frac{K'_{cW}(Alk - HCO_3^-)}{(K'_{c2}.HCO_3^- + K'_{cW})} + \frac{HCO_3^-(K'_{c2}.HCO_3^- + K'_{cW})}{K'_{c1}(Alk - HCO_3^-)} + HCO_3^-$$

$$+ \frac{(K'_{c2}.HCO_3^- + K'_{cW})}{(Alk - HCO_3^-)} = 0 \qquad (65)$$

Substituting Eqs. (64 and 60) into Eq. (49) gives:

$$C_2 = Alk - \frac{K'_{cS}}{K'_{c2}.HCO_3^-} \cdot \frac{(K'_{c2}.HCO_3^- + K'_{cW})}{(Alk - HCO_3^-)} \qquad (66)$$

Equation (65) is a cubic equation in HCO_3^-. For a fixed chosen value of Alkalinity, HCO_3^- is solved for a series of values of C_1. The C_2 value corresponding to each C_1 value is calculated from Eq.(66) using the relevant HCO_3^- value from Eq. (65). Plotting the corresponding C_1 and C_2 values traces the vertical section of an Alkalinity line in the Modified Caldwell-Lawrence diagram.

The complete Alkalinity line for some chosen Alkalinity value is plotted in the diagram as follows: The lower horizontal limb of the Alkalinity line is first plotted using Eqs. (62 and 63) until the slope in the diagram of this line is greater than 45°. The vertical section of the Alkalinity line is then plotted using Eqs. (65 and 66) until the slope of this line is less than 45°. The equations for the horizontal section of the Alkalinity line are then again used to trace the upper horizontal limb of the Alkalinity line.

In Appendix C a selection of Modified Caldwell-Lawrence diagrams is given for ranges of Ca, Alk, pH, ionic strength and temperature likely to include most natural waters and their'transient' conditions during water treatment. In Appendix D is a print-out of a computer program to plot the Modified Caldwell-Lawrence diagram for any selected scale, diagram limits, ionic strength and temperature.

The equations for the co-ordinate parameters C_1 and C_2, Eqs. (48 and 49) differ from those of Caldwell and Lawrence (1952) who apparently omitted the hydrolyses effect of water in their derivations. The consequence of this omission is that all their subsequent equations incorporating these two parameters are in error, though this error may be insignificant and negligible for practical purposes. However, a more serious consequence is that it is not possible with their equations to express in a consistent and logical manner the effect of addition of chemicals to an under- or oversaturated water.

In effect the procedures described by Caldwell and Lawrence for estimating the $CaCO_3$ saturated condition of any water are correct but cannot be logically derived from their co-ordinate parameters.

APPLICATION OF THE MODIFIED CALDWELL-LAWRENCE DIAGRAM IN WATER TREATMENT

The Modified Caldwell-Lawrence diagram can be used for solving a number of water treatment problems including water stabilization and softening. These two problems are of particular importance because of their general application in water treatment and are thus given special attention in the text. The theoretical basis and techniques for graphically solving each of these types of problems are discussed in detail below.

Application to Stabilization Problems

Stabilization problems generally entail calculating chemical dosage requirements to adjust Alkalinity and pH of a water to some values which make a water well buffered against pH change and oversaturated with respect to $CaCO_3$. Calculations are carried out in the diagram in two steps. The first step entails adjusting the parameters Alkalinity, pH and Acidity (and perhaps Ca^{++})

considering only equilibria between dissolved species, i.e. $CaCO_3$ is considered infinitely soluble. The second step entails estimating the $CaCO_3$ saturation state of the water adjusted in the first step, calculating the mass of $CaCO_3$ which could potentially precipitate or dissolve into the water and estimating the final saturated equilibrium state which such a water would ultimately attain.

Thus, in effect, calculations for estimating chemical dosage requirements in stabilization problems consider $CaCO_3$ saturation and subsequent precipitations a result of changes which have already occurred in the aqueous phase.

A number of examples demonstrating the application of these concepts in the Modified Caldwell-Lawrence diagram are given below:

Example

30 ppm Na_2CO_3 expressed as $CaCO_3$ is added to CO_2 free distilled water. What is the pH, Alkalinity, Acidity and total carbonic species concentration of the water?

Initial Alkalinity and Acidity of neutral distilled water is zero.

$$Alk_{added} = CO_{3\ added}^{=} + HCO_{3\ added} + OH_{added} - H_{added}$$

$$= 30 + 0 + 0 - 0$$

$$= 30 \text{ ppm as } CaCO_3$$

i.e.

Final Alkalinity = 0 + 30 = 30 ppm

From Eq. (21):

$$Acidity_{added} = CO_{2\ added} + HCO_{3\ added} + H_{added} - OH_{added}$$

$$= 0$$

i.e.

Final Acidity = 0

From Eq. (54), Chapter 3:

Total carbonic species = C_T = (30 + 0)/2 = 15 ppm as $CaCO_3$
        concentration

Identifying Acidity = 0 and Alkalinity = 30 ppm lines in
Figure 16, their intersection point gives pH = 10,15.

Example
----

60 ppm of $NaHCO_3$ expressed as $CaCO_3$ is added to CO_2 free distilled water. What is the pH, Alkalinity, total carbonic species
concentration and Acidity?

   From Eq. (20):

Alk_{added} = $CO_3^=$ added + HCO_3 added + OH_{added} − H_{added}

           = 0 + 60 + 0 − 0

           = 60 ppm as $CaCO_3$
i.e.

Final Alkalinity = (0 + 60) = 60 ppm as $CaCO_3$

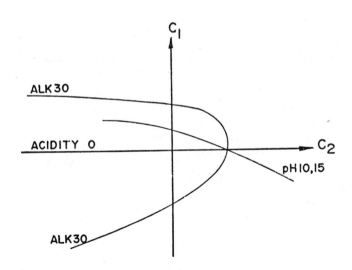

Figure 16.   30 ppm Na_2CO_3 is added to CO_2 free water

From Eq. (21):

$$\text{Acidity}_{added} = CO_{2\;added} + HCO_{3\;added} + H_{added} - OH_{added}$$

$$= 0 + 60 + 0 - 0$$

$$= 60 \text{ ppm as } CaCO_3$$

i.e.

Final Acidity = $(0 + 60)$ = 60 ppm as $CaCO_3$

From Eq. (54), Chapter 3:

Total carbonic species = C_T = $(60 + 60)/2$ = 60 ppm as $CaCO_3$
    concentration

In the modified Caldwell-Lawrence diagram draw the Acidity ordinate line 60 ppm as $CaCO_3$ and identify Alkalinity line 60 ppm. Their intersection, Point 1, Figure 17, gives pH 8,1.

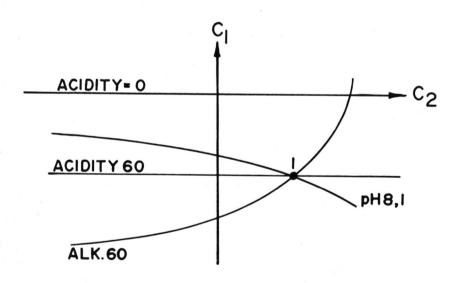

Figure 17. 60 ppm $NaHCO_3$ is added to CO_2 free water

Example

30 ppm Na_2CO_3 and 30 ppm $NaHCO_3$ are added to distilled water containing 2 ppm CO_2 (as $CaCO_3$). What is the final Alkalinity, Acidity, total carbonic species concentration and pH?

Initial Alkalinity = 0

From Eq. (21):

Initial Acidity = $CO_{2\ added}$ + $HCO_{3\ added}$ + H_{added} - OH_{added}

$$= 2 + 0 + 0 - 0 = 2\ \text{ppm as } CaCO_3$$

Initial total carbonic = (0 + 2)/2 = 1 ppm as $CaCO_3$
species concentration

Addition of chemical dosage:

From Eq. (20):

Alk_{change} = $CO_{3\ added}$ + $HCO_{3\ added}$ + OH_{added} - H_{added}

$$= 30 + 30 + 0 - 0$$

$$= 60\ \text{ppm as } CaCO_3$$

From Eq. (21):

$Acidity_{change}$ = $CO_{2\ added}$ + $HCO_{3\ added}$ + H_{added} - OH_{added}

$$= 0 + 30 + 0 - 0$$

$$= 30\ \text{ppm as } CaCO_3$$

Final Alkalinity = (0 + 60) = 60 ppm as $CaCO_3$

Final Acidity = (30 + 2) = 32 ppm as $CaCO_3$

Final total carbonic = (60 + 32)/2 = 46 ppm as $CaCO_3$
species concentration

Intersection of Alkalinity line 60 ppm and Acidity line 32 ppm, Point 1 in Figure 18 gives final pH of 9,7.

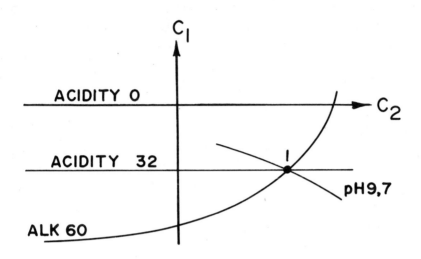

Figure 18.   30 ppm Na_2CO_3 and 30 ppm $NaHCO_3$ are added to water
containing 2 ppm CO_2 (as ppm $CaCO_3$)

Example

Water has an Alkalinity of 40 ppm as $CaCO_3$, pH 8,5 and no calcium,
20 ppm of NaOH (as ppm $CaCO_3$) is added. What is the Alkalinity,
Acidity, total carbonic species concentration and pH?

Identify Alkalinity = 40 and pH = 8,5 lines (Figure 19). Intersection (Point 1) defines the Acidity, 38 ppm as $CaCO_3$. Adding
20 ppm NaOH (as ppm $CaCO_3$) has the following effect:

From Eq.(20):

$$Alk_{added} = CO_{3\ added} + HCO_{3\ added} + OH_{added} - H_{added}$$

$$= 0 + 0 + 20 - 0 = 20 \text{ ppm as } CaCO_3$$

From Eq. (21):

$$Acidity_{added} = CO_{2\ added} + HCO_{3\ added} + H_{added} - OH_{added}$$

$$= 0 + 0 + 0 - 20 = -20 \text{ ppm as } CaCO_3.$$

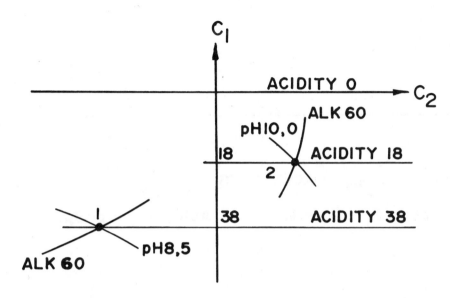

Figure 19. 20 ppm NaOH is added to water initially with zero Ca, 40 ppm Alkalinity and pH 8,5

i.e. there is a decrease in Acidity.

New Alkalinity = $(40 + 20)$ = 60 ppm as $CaCO_3$

New Acidity   = $(38 - 20)$ = 18 ppm as $CaCO_3$

From Eq. (54), Chapter 3:

Total carbonic species = C_T = $(60 + 18)/2$ = 39 ppm as $CaCO_3$
       concentration

   Intersection of Alkalinity 60 and Acidity 18 lines, Point 2, Figure 19, gives new pH as 10,0.

Example

A water has an Alkalinity of 40 ppm as $CaCO_3$, pH 8,5 and zero calcium. 20 ppm of Na_2CO_3 (as ppm $CaCO_3$) is added to the water. What is the Alkalinity, Acidity, total carbonic species concentration and pH?

   Identify the Alkalinity 40 and pH 8,5 lines (Figure 20).

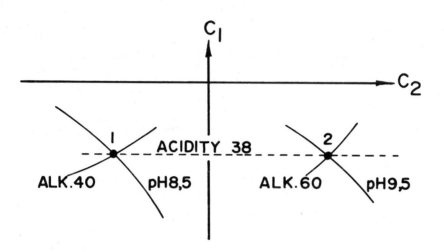

Figure 20. 20 ppm Na_2CO_3 is added to water initially with zero Ca, 40 ppm Alkalinity and pH 8,5

Intersection (Point 1) defines Acidity (38 ppm). Addition of Na_2CO_3 has the following effect:

From Eq. (20):

$$Alk_{change} = CO_{3\ added} + HCO_{3\ added} + OH_{added} - H_{added}$$

$$= 20 + 0 + 0 - 0$$

$$= 20 \text{ ppm as } CaCO_3$$

From Eq. (21):

$$Acidity_{change} = CO_{2\ added} + HCO_{3\ added} + H_{added} - OH_{added}$$

$$= 0 + 0 + 0 - 0 = 0$$

i.e.

New Alkalinity = (20 + 40) = 60 ppm as $CaCO_3$

New Acidity   = as before, i.e. 38 ppm as $CaCO_3$

From Eq. (54), Chapter 3:

Total carbonic species = $(60 + 38)/2 = 49$ ppm as $CaCO_3$
      concentration

Intersection of new Alkalinity and Acidity lines, Point 2, Figure 20, defines new pH 9,5.

Example

Analysis of a water gives Alkalinity 30 ppm, Ca^{++} 15 ppm (both as ppm $CaCO_3$) and pH 8,9. What is the state of the water after adding 10 ppm $Ca(OH)_2$ (as ppm $CaCO_3$)?

The initial ionic equilibrium state between dissolved species in the water is given in the diagram by the intersection point of pH 8,9 and Alkalinity 30 ppm line, Point 1 in Figure 21. Acidity ordinate value of Point 1 defines the initial Acidity of the water, i.e. 26 ppm as $CaCO_3$

The initial $CaCO_3$ saturation state of the water is found as

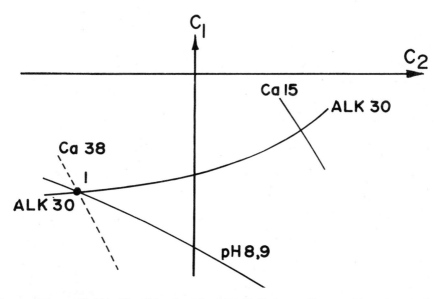

Figure 21. Alkalinity 30, Ca 15 and pH 8,9 - water undersaturated with respect to $CaCO_3$

follows: the value of the calcium line through Point 1 defines the Ca^{++} concentration required for saturation, i.e. 38 ppm as $CaCO_3$. The actual Ca^{++} concentration is only 15 ppm as $CaCO_3$. Comparing the required Ca^{++} value with the actual value one deduces that the water is undersaturated with respect to $CaCO_3$.

The addition of 10 ppm $Ca(OH)_2$ to this water may cause the solubility product for $CaCO_3$ to be exceeded with subsequent precipitation. For this reason calculations are now executed in two steps: (i) the new state of the water is found assuming $CaCO_3$ to be infinitely soluble; (ii) the $CaCO_3$ saturation state is assessed - if oversaturated the final saturated equilibrium state and the mass of $CaCO_3$ precipitated are calculated.

(i) Adding 10 ppm $Ca(OH)_2$ (as ppm $CaCO_3$) and assuming $CaCO_3$ as infinitely soluble:

From Table 2 the Ca^{++} concentration increases by 10 ppm as $CaCO_3$, i.e.

$$Ca^{++} = (15 + 10) = 25 \text{ ppm as } CaCO_3$$

From Eq. (20):

$$Alk_{added} = HCO^-_{3 \text{ added}} + H_2CO_{3 \text{ added}} + OH^-_{added} - H^+_{added}$$

$$= 0 + 0 + 10 - 0$$

$$= 10 \text{ ppm as } CaCO_3$$

Final Alkalinity = $(30 + 10)$ = 40 ppm as $CaCO_3$

From Eq. (21):

$$Acidity_{added} = HCO^-_{3 \text{ added}} + H_2CO^*_{3 \text{ added}} + H^+_{added} - OH^-_{added}$$

$$= 0 + 0 + 0 - 10$$

$$= - 10 \text{ ppm as } CaCO_3$$

Final Acidity = $(26 - 10)$ = 16 ppm as $CaCO_3$

pH is defined by the value of the pH line through the
intersection point of lines representing Alkalinity
40 ppm and Acidity 16 ppm, i.e. pH 9,8 through Point 3,
Figure 22.

(ii) The $CaCO_3$ saturation state of the water is now assessed.

Calcium value required for saturation is given by the
value of the calcium line through Point 3, Figure 22,
i.e. Ca^{++} = 8 ppm as $CaCO_3$. Comparing this required
calcium value with the actual calcium concentration one
deduces the water is oversaturated and $CaCO_3$ precipitates with time. The saturated state which is ultimately
achieved is found in the Caldwell-Lawrence diagram as
follows:

Two conditions must be satisfied when $CaCO_3$ precipitates
from an oversaturated water:

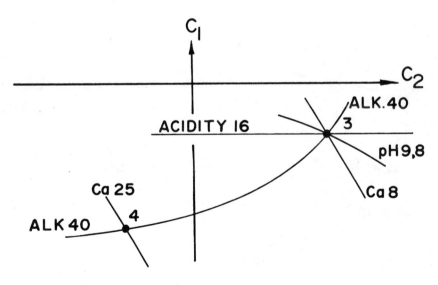

Figure 22. 10 ppm $Ca(OH)_2$ is added to water initially with Ca 15,
Alkalinity 30 (both ppm as $CaCO_3$) and pH 8,9

(a) From Eq. (38) during precipitation of $CaCO_3$ the Ca^{++} and $CO_3^=$ change by equal amounts, i.e. the value of the parameter (Alk-Ca) remains constant. Thus the final saturated condition lies on the (Alk-Ca) line for the oversaturated water, i.e. on the vertical line through the intersection point of the lines representing the oversaturated values of Alkalinity and calcium - for the problem in question, on the vertical line through Point 4, Figures 22 and 23.

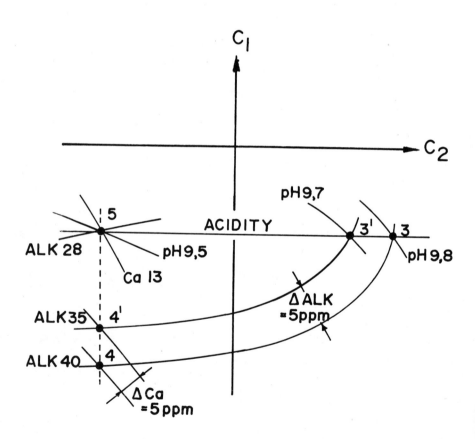

Figure 23. $CaCO_3$ precipitates from an oversaturated water.

(b) From Eq. (21), abstraction of $CO_3^=$ during precipi-
     tation does not affect the Acidity, i.e. Acidity
     remains constant. Thus the saturated state ulti-
     mately achieved lies on the Acidity line represent-
     ing the initial oversaturated Acidity value, i.e.
     on the horizontal line through the intersection
     point of the lines representing the oversaturated
     values of Alkalinity and pH - for the problem in
     question, on the horizontal line through Point 3,
     Figure 22 and 23.

Intersection point of these two lines defines the final saturated equilibrium state of the water in the diagram, Point
5, Figure 23. The values of the calcium, Alkalinity and
pH lines through this point are those values which the water
ultimately achieves, i.e. Alkalinity 28 ppm, calcium 13 ppm
(both as $CaCO_3$) and pH 9,5.

The mass of $CaCO_3$ which ultimately precipitates from the
water is given by either the change in calcium or Alkalinity, i.e.

$CaCO_3$ precipitated = Δ Alk

$$= (40 - 28) = 12 \text{ ppm as } CaCO_3$$

$$= \Delta \text{ Ca}$$

$$= (25 - 13) = 12 \text{ ppm as } CaCO_3$$

Using the reasoning developed above, and the Modified Caldwell-
Lawrence diagram, one can trace the transient states of an oversaturated water as $CaCO_3$ precipitates. For example, in the above
problem after 5 ppm $CaCO_3$ has precipitated:

From Table 2

Calcium concentration = $(25 - 5) = 20$ ppm as $CaCO_3$

Alkalinity $= (40 - 5) = 35$ ppm as $CaCO_3$

No change occurs in Acidity, i.e.

Acidity = 16 ppm as $CaCO_3$

Intersection of the lines representing calcium 20 ppm and Alkalinity 35 ppm occurs at Point 4' (vertically above Point 4) in Figure 23. The new state of equilibrium between the dissolved carbonic species is given by Point 3' (with the same Acidity as Points 3 and 5) the intersection of the lines representing Alkalinity 35 ppm and Acidity 16 ppm. Point 3', Figure 23, also defines the new pH in the water, i.e. pH 9,7.

Example

Excess solid $CaCO_3$ is added to CO_2 free distilled water. Determine the concentration of $CaCO_3$ which dissolves into the water. (Temperature of the water is 25°C).

Initial Alkalinity, Acidity and calcium concentrations of neutral distilled water are zero.

When $CaCO_3$ dissolves into water neither Acidity not (Alk-Ca) change (see Eqs. (21 and 41)). This forms the basis for the solution.

For CO_2 free water the initial state of the water is Acidity = 0, Alkalinity = 0, Ca^{++} = 0 (all as ppm $CaCO_3$). As $CaCO_3$ dissolves the concentration Ca^{++} and Alkalinity increase by equal amounts. Acidity does not change. Thus the final state is:

Acidity = 0; (Alk-Ca) = 0 (both as ppm $CaCO_3$)

Using the chart in Figure 14b (ionic strength = 0,001, temperature = 25°C) intersection of the co-ordinates Acidity = 0 and (Alk-Ca) = 0 defines the final equilibrium state (Point 1), i.e.

Ca^{++} = 14 ppm, Alkalinity = 14 ppm and pH = 9,95

Example

A water has an Alkalinity = 40 ppm and Ca^{++} = 10 ppm (both as $CaCO_3$), pH = 8,5 and ionic strength = 0,001. Excess $CaCO_3$ is

added to the water. Determine the mass of $CaCO_3$ dissolving and
the final constitution of the water.

Initial Acidity is given by the intersection of the pH = 8,5
line and the Alk = 40 line. Hence initial constitution is:

Acidity = 39; (Alk-Ca) = (40 - 10) = 30   (Figure 14b, Point 2)

Addition of $CaCO_3$ does not change Acidity and (Alk-Ca) (see
example above). Hence final condition is:

Acidity = 39; (Alk-Ca) = 30

and the water is saturated. The solution is the intersection of
the (Alk-Ca) and Acidity co-ordinates, i.e. (Figure 14b, Point 3)

Ca^{++} = 14;   Alk = 44; Acidity = 39 (all as ppm $CaCO_3$); pH = 9,0

ΔCa = 14 - 10 = 4; ΔAlk = 44 - 40 = 4

$\Delta CaCO_3$ = ΔCa = ΔAlk = 4 ppm as $CaCO_3$

Example

Analysis of a water gives: Alkalinity 35 ppm, calcium 50 ppm
(both as $CaCO_3$) and pH 10,8. Calculate the state of the water
after each of the following chemicals are added: (i) 60 ppm CO_2;
(ii) 68 ppm CO_2; (iii) 72 ppm CO_2 (all as ppm $CaCO_3$).

(i) Plotting the lines representing the measured values of
     Alkalinity, calcium and pH in the Modified Caldwell-
     Lawrence diagram - they all intersect at Point 1,
     Figure 24, i.e. the water is just saturated with res-
     pect to $CaCO_3$. Acidity of the water is given by the
     Acidity ordinate value of Point 1. i.e. Acidity =
     - 35 ppm as $CaCO_3$. Adding 60 ppm of CO_2 (as $CaCO_3$)
     has the following effect on the water:

     Initially considering only equilibria between dis-
     solved species, from Eq. (20):

Alkalinity change = HCO_3^- added $+ CO_3^=$ added $+ OH^-$ added $- H^+$ added

$$= 0 + 0 + 0 - 0$$

$$= 0$$

i.e. Alkalinity does not change with CO_2 addition *provided* $CaCO_3$ is not precipitated.

From Eq. (21):

Acidity change = $H_2CO_3^*$ added $+ HCO_3^-$ added $+ H^+$ added $- OH^-$ added

$$= 60 + 0 + 0 - 0$$

$$= 60 \text{ ppm as } CaCO_3 \quad \text{i.e.}$$

New Acidity = $- 34 + 60$

$$= + 26 \text{ ppm as } CaCO_3$$

The new pH in the water is given by the value of the pH line through the intersection point of the lines representing the new Acidity and Alkalinity values, i.e. pH 9,4 line through Point 2, Figure 24.

The Ca^{++} concentration remains at 50 ppm as $CaCO_3$ on the assumption that no precipitation has occurred.

To ascertain the saturation state of the water the value of the calcium line through Point 2 (i.e. Ca^{++} 20 ppm) is compared with the actual calcium value (50 ppm) - one deduces that the water is oversaturated with respect to $CaCO_3$.

The mass of $CaCO_3$ which ultimately precipitates from the water to achieve a saturated state is now calculated using the same technique as in the example above.

Neither of the parameters Acidity nor (Alk-Ca) changes

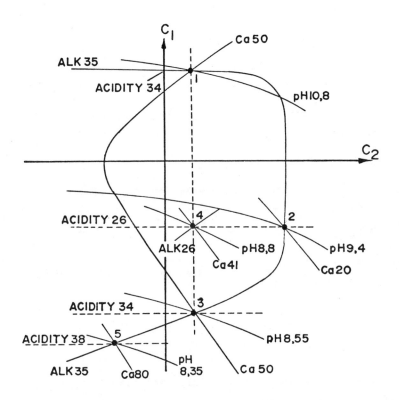

Figure 24. Addition of 60 or 68 or 72 ppm CO_2 (as ppm $CaCO_3$) to water with Ca 15, Alkalinity 35 and pH 10,8

with $CaCO_3$ precipitation. The final saturated state is given by the intersection of the Acidity (or horizontal) line through Point 2 and the vertical line through Point 3, i.e. Point 4, Figure 24, with Alkalinity 26 ppm, calcium 41 ppm (both as ppm $CaCO_3$) and pH 8,8.

The mass of $CaCO_3$ precipitated is given by either the

Ca^{++} or Alkalinity precipitated, i.e.

$CaCO_3$ precipitated = ΔAlk = ΔCa

$$= (35 - 26) = (50 - 41)$$

$$= 9 \text{ ppm}$$

(ii) Initial state of the water plots at Point 1 in Figure 24 and is just saturated with respect to $CaCO_3$.

Adding 68 ppm CO_2 (as $CaCO_3$) to the water has the following effect:

From Eq. 20):

No change occurs in the Alkalinity value of the water provided no $CaCO_3$ precipitation occurs.

From Eq. (21):

Acidity change = HCO_3^* added + HCO_3^- added + H_{added}^+ - OH_{added}^-

$$= 68 + 0 + 0 - 0$$

$$= 68 \text{ ppm as } CaCO_3 \qquad \text{i.e.}$$

New Acidity = $- 34 + 68 = + 34$ ppm as $CaCO_3$

Intersection of the new Acidity and Alkalinity values Point 3 defines the new pH in the water, i.e. pH 8,55

Assuming no $CaCO_3$ precipitation occurs the line representing the original Ca^{++} value, 50 ppm as $CaCO_3$, passes through Point 3, i.e. the water is just saturated with respect to $CaCO_3$.

(iii) Adding 72 ppm CO_2 (as $CaCO_3$) to a water whose condition plots at Point 1 in Figure 24 - a similar line of reasoning as used in the examples above gives the new state of the water:

From Eq. (20) the new Alkalinity value is equal
to the original Alkalinity value.

Acidity change = + 72 ppm as $CaCO_3$

New Acidity   = - 34 + 72

              = 38 ppm as $CaCO_3$

Intersection of the lines representing the new
Acidity and Alkalinity values occurs at Point 5,
Figure 24. The value of the pH line through Point
5 defines the new pH in the water, i.e. pH 8,35.

The value of the calcium line passing through Point
5 gives the calcium concentration required for
saturation, i.e. Ca^{++} 80 ppm as $CaCO_3$. Comparing
this required calcium value with the actual calcium
concentration in the water (50 ppm as $CaCO_3$) one
deduces that the water is now undersaturated with
respect to $CaCO_3$.

Interpretation of the Langelier Saturation Index

In present day water treatment practice the Langelier Saturation
Index is the parameter generally used to define the $CaCO_3$ saturation state. The Index, and various deficiencies associated
with its practical interpretation, were discussed in depth in
Chapter 4. The examples given in this section show that the
Modified Caldwell-Lawrence diagram is easily used to give not
only the $CaCO_3$ saturation state of a water but also a quantitative
measure of the mass of $CaCO_3$ which the water will ultimately precipitate or dissolve. Thus by using the conditioning diagram the
$CaCO_3$ saturation state can be reported in a qualitative and quantitative sense. This data is much more meaningful than reporting
the Saturation Index value. However, because the Saturation Index is widely used in present day water treatment practice, it is

important to illustrate how the Index value can be estimated graphically in the Modified Caldwell-Lawrence diagram and also to
illustrate graphically the various deficiencies associated with
its practical interpretation.

The Langelier Saturation Index was defined by Eq. (9), Chapter
4, i.e.

$$S.I. = pH - pH_s$$

where pH_s is the theoretical pH value which gives saturation for
the measured calcium and Alkalinity values in a water. Langelier
showed that the value of pH_s can be calculated (see page 132).The
saturation state of a water is then given by the sign of the Saturation Index, i.e. S.I. positive indicates oversaturation and
negative undersaturation.

In the Modified Caldwell-Lawrence diagram pH_s is given by the
value of the pH line through the intersection point of the lines
representing the measured Alkalinity and calcium values in the
diagram.

Example

Analysis of a water gives Alkalinity 40 ppm, calcium 30 ppm (both
as $CaCO_3$) and pH 9,6. Estimate the Langelier Saturation Index in
the Modified Caldwell-Lawrence diagram.

Lines representing the measured Alkalinity and calcium values
intersect in Point 1, Figure 25. The value of the pH line through
Point 1 is pH 8,95, i.e. pH_s = 8,95 and

$$S.I. = pH - pH_s$$

$$= 9,6 - 8,95 = + 0,65$$

The positive sign indicates oversaturation with respect to
$CaCO_3$.

A number of important points regarding interpretation of the
Langelier Saturation Index can now be illustrated by graphical

means in the Modified Caldwell-Lawrence diagram:

(i) The final saturated pH which an over- or undersaturated
     water ultimately achieves after precipitating or dis-
     solving $CaCO_3$ is different from the pH_s values used to
     calculate S.I.

Example

Analysis of a water gives Alkalinity 40 ppm, calcium
30 ppm (both as $CaCO_3$) and pH 9,6. Plotting the state
of the water in Figure 26, Langelier's pH_s, is given
by the value of the pH line through the intersection
point of the Alkalinity 40 ppm and Ca 30 ppm lines,
i.e. Point 1 with pH_s = 8,95.

The final saturated state which the water ultimately
achieves is given by the intersection of the initial
Acidity and (Alk-Ca) lines, i.e. Point 2, Figure 26.

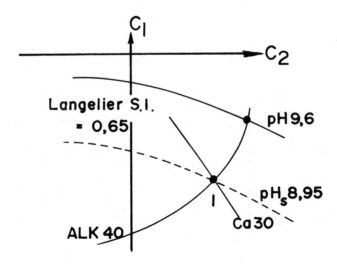

Figure 25. Estimate of the Langelier Saturation Index

The pH line through Point 2 gives the actual saturated
pH value which the water ultimately achieves, i.e.
pH 9,2.

(ii) The significance of the Langelier Index is in its *sign*
      (either positive or negative). The magnitude of the
      Index has no significance and certainly does not indi-
      cate how much $CaCO_3$ will eventually either precipitate
      or dissolve into the water as it achieves saturation.
      Two waters may have the same saturation index and dis-
      solve (or precipitate) very different concentrations of
      $CaCO_3$ to reach saturation. For example, the condition
      of two waters (A) and (B) is:

Water (A): Alkalinity 40 ppm, Ca 30 ppm (both as $CaCO_3$), pH 9,6

Water (B): Alkalinity 50 ppm, Ca 70 ppm (both as $CaCO_3$), pH 9,1

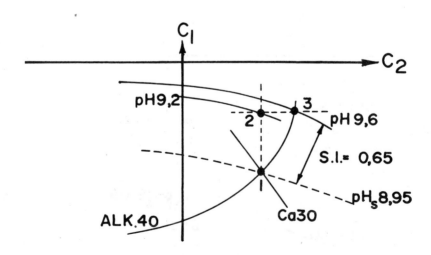

Figure 26. Estimation of the true supersaturation state

Plotting the state of each water in Figure 27 shows
that both waters have a Langelier saturation index
of + 0,65. Graphical techniques establish the final
saturation condition for each water as:

Water (A): Alkalinity 33 ppm, Ca 23 ppm (both as $CaCO_3$), pH 9,2

Water (B): Alkalinity 46 ppm, Ca 66 ppm (both as $CaCO_3$), pH 8,53

Concentration of $CaCO_3$ precipitated from each water is
given by the difference between the initial oversaturated value of calcium (or Alkalinity) and the final saturated value of calcium (or Alkalinity). Thus, though
both waters have a Langelier saturation index of + 0,65
whereas 7 ppm of $CaCO_3$ is precipitated from water (A)
only 4 ppm of $CaCO_3$ is precipitated from water (B).

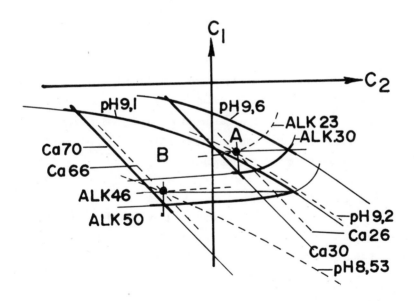

Figure 27. Two waters with equal S.I. values precipitate different masses of $CaCO_3$ to saturation

(iii) The Langelier Saturation Index reverses for pH above
       the carbonate equivalence point. Increasing the pH
       of a saturated water which plots *above* the zero hori-
       zontal axis creates undersaturation, while for a sa-
       turated water plotting *below* the horizontal axis,
       increasing pH creates oversaturation. Thus the
       Langelier saturation equation, S.I. = pH - pH_s, *must
       be rewritten as follows for those waters whose con-
       dition plots above the horizontal axis:*

S.I. = pH_s - pH

     where all the terms have been previously defined.

     A positive index, as below, indicates oversaturation
     and a negative index undersaturation.

     This reversal in the Saturation Index equation is
     explained by noting that the zero horizontal axis is
     a line describing *an equivalent carbonate solution,*
     (i.e. $OH^- = H^+ + HCO_3^- + H_2CO_3^*$). Water whose condition
     plots above this axis is almost completely governed
     by hydroxide ions and CO_2 must be added to *lower the
     pH* to obtain $CaCO_3$ supersaturation. 'Normal' water
     plots below the horizontal axis and pH *must be raised*
     to bring about *oversaturation.*

Application to Water Softening

In using the Modified Caldwell-Lawrence diagram for water softening calculations it is convenient to consider the treated water
as always saturated with respect to $CaCO_3$, i.e. calculations are
carried out assuming two phase equilibrium - equilibrium between
ionic species and solid $CaCO_3$. For dosage calculation purposes
in water softening only the final saturated state that a water
achieves with time is considered relevent.

Lines representing pH, Alkalinity and calcium values of a $CaCO_3$ saturated water intersect in a point in the diagram. Adding to the water any of the chemicals commonly used in water treatment practice (see Table 1) causes $CaCO_3$ either to precipitate or dissolve as a new saturated equilibrium condition is established - defined by a new point in the diagram. The position in the diagram of the new saturated equilibrium point relative to the initial point is described by a vector of magnitude and direction depending on the mass and type of chemical added. For rapid utilization of the diagram to adjust a water from one saturated state to another, it is convenient to develop a 'direction format diagram'. The format diagram gives the direction which a point representing a saturated water in the conditioning diagram moves as a result of adding a specific type of chemical dose. (See Figure 28). This 'direction format diagram' is developed as follows"

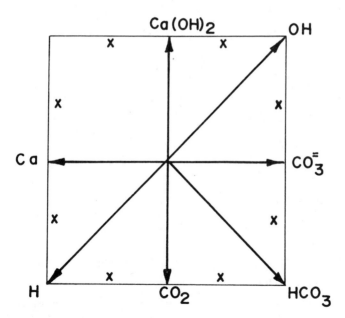

Figure 28. Direction Format Diagram for a saturated water in contact with solid $CaCO_3$

1. <u>Adding Ca^{++} to a saturated water</u>

Adding calcium to a saturated water causes oversaturation and precipitation of $CaCO_3$. Eventually the water achieves a new state of saturated equilibrium. Two effects are noted:

(i) From Eq. (21) loss of $CO_3^=$ as $CaCO_3$ precipitates does not change the Acidity.

(ii) A mass balance of the effect of adding Ca^{++} gives:

Ca^{++}_{added} = measured increase in Ca^{++} concentration
$\qquad$ + concentration of Ca^{++} precipitated as $CaCO_3$

$\qquad$ = ΔCa + Ca precipitated $\qquad\qquad\qquad\qquad$ (67)

Δ denotes an increase.

$\quad$ From Eq. (38):

Ca precipitated = $CO_3^=$ precipitated $\qquad\qquad\qquad\qquad$ (68)

$\quad$ Substituting Eq. (68) into Eq. (67):

Ca^{++}_{added} = ΔCa + $CO_3^=$ precipitated $\qquad\qquad\qquad$ (69)

$\quad$ and from Eq. (20):

$CO_3^=$ precipitated = Δ Alk $\qquad\qquad\qquad\qquad\qquad$ (70)

$\quad$ Substituting Eq. (70) into Eq. (69):

Ca^{++}_{added} = Δ Ca - Δ Alk

$\qquad$ = Δ(Ca - Alk)

$\qquad$ = - Δ C_2

$\quad$ Thus, adding calcium to a saturated water moves the point describing the water in the Caldwell-Lawrence diagram horizontally to the left a distance equal to the concentration of calcium added as ppm $CaCO_3$.

Example

X ppm of Ca^{++} as $CaCO_3$ is added to a saturated water whose initial condition (defined by pH, Alk and Ca) plots at Point 1, Figure 29. $CaCO_3$ precipitates from the water and the following conditions must be satisfied:

(i)   The Acidity of the water does not change, i.e. the new condition of the water in the diagram lies on the horizontal line through Point 1.

(ii) As shown above, the C_2 ordinate value changes as follows:

$$Ca^{++}_{added} = X = - \Delta C_2$$

The new condition of the water is given in the diagram by the values of pH, Alk and Ca lines passing through Point 2 in Figure 29.

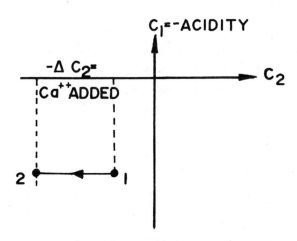

Figure 29. Ca added to a saturated water

2. <u>Addition of $CO_3^=$</u>

Adding $CO_3^=$ to a saturated water causes oversaturation and $CaCO_3$ precipitation. After the water has again achieved saturation the following effects are noted:

(i) From Eq. (21) Acidity of the water does not change due to either addition of $CO_3^=$ or loss of $CO_3^=$ in the $CaCO_3$ precipitate.

(ii) A mass balance equation resulting from $CO_3^=$ addition and subsequent $CaCO_3$ precipitate is:

$CO_3^=$ added = increase in Alkalinity of the water

$\qquad\qquad + CO_3^=$ precipitated as $CaCO_3$

$\qquad\quad = \Delta$ Alk $+ CO_3^=$ precipitated

Δ denotes an increase

Also equal concentrations of $CO_3^=$ and Ca^{++} precipitate and the above equation becomes:

$CO_3^=$ added $= \Delta$ Alk + Ca precipitated

$\qquad\quad = \Delta$ Alk $- \Delta$ Ca

$\qquad\quad = \Delta C_2$ $\qquad\qquad\qquad\qquad\qquad\qquad\qquad$ (71)

Thus, adding $CO_3^=$ to a saturated water moves the point representing the water in the Modified Caldwell-Lawrence diagram to the right a distance equal to the concentration of $CO_3^=$ added expressed as ppm $CaCO_3$.

<u>Example</u>

Y ppm of $CO_3^=$ (expressed as $CaCO_3$) is added to a saturated water whose condition plots at Point 1 in Figure 30. $CaCO_3$ is precipitated and the following conditions must be satisfied:

(i) Acidity in the water does not change and the final saturated condition lies on the horizontal line through Point 1.

(ii) From Eq. (71):

$$CO_3^= \text{ added} = \Delta C_2 \quad \text{i.e.}$$

$$\Delta C_2 = Y \text{ ppm as } CaCO_3$$

The new state of the water is given by the values of pH, Alkalinity and Ca lines through Point 2, Figure 30.

3. Addition of CO_2

Adding CO_2 to a saturated water in contact with $CaCO_3$ has the following effects:

(i) From Eq. (21) CO_2 added to a water increases Acidity
     by an equivalent amount, i.e.

$$CO_2 \text{ added} = - \Delta C_1$$

(ii) If CO_2 is added to a saturated water *not* in contact
      with $CaCO_3$, Alkalinity does not change, pH, however,
      decreases. The decrease in pH causes undersaturation.

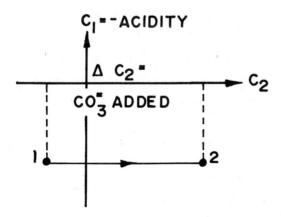

Figure 30. $CO_3^=$ added to a saturated water

Should the water be in contact with $CaCO_3$, the $CaCO_3$
dissolves with time until saturation is achieved.
Thus, adding CO_2 to a water in contact with $CaCO_3$
causes equal concentrations of carbonate and calcium
to dissolve, i.e.

$$\Delta \ Ca = CO_3^= \ dissolving$$

where Δ denotes an increase.
From Eq. (20):

$$CO_3^= \ dissolving = \Delta \ Alk$$

i.e. $\Delta \ Ca^{++} \ \ = \Delta \ Alk$

i.e. $\Delta(Alk\text{-}Ca) = 0$

i.e. $\Delta \ C_2 \ \ \ \ = 0$

Thus, the effect of adding CO_2 to a saturated water in contact
with solid $CaCO_3$ is to move the point defining the water in the
Modified Caldwell-Lawrence diagram vertically downwards by a distance equal to the concentration of CO_2 added, expressed as $CaCO_3$.

Example

Z ppm CO_2 as $CaCO_3$ is added to a saturated water in contact with
$CaCO_3$. The initial condition plots at Point 1, in conditioning
diagram Figure 31.

Adding CO_2 to a water in contact with $CaCO_3$ does not change
the C_2 ordinate value of the water in the diagram, i.e. the final
condition of the water lies on the vertical line through Point 1
Figure 31.

Acidity of the water increases by an amount equal to the concentration of CO_2 added, i.e.

$$CO_2 \ added = -\ \Delta \ C_1$$

The new state of the water is given by pH, Alk and Ca lines
through Point 2 in Figure 31.

4. Addition of Ca(OH)$_2$

Adding $Ca(OH)_2$ to a saturated water has the following effects:

(i) Alkalinity changes. Equal concentrations of calcium
     and hydroxide are added, i.e.

$$Ca^{++}_{added} = OH^-_{added}$$

and from Eq. (20):

$$OH^-_{added} = Alk_{added} \quad i.e.$$

$$Ca^{++}_{added} = Alk_{added}$$

Furthermore, as a result of OH^- addition the water becomes
oversaturated and $CaCO_3$ precipitates, i.e.

$$Ca^{++} \; precipitated = CO_3^= \; precipitated$$

$$= Alk \; precipitated$$

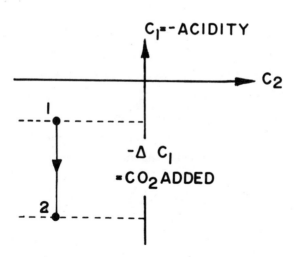

Figure 31. CO_2 added to a saturated water in contact with solid
$CaCO_3$

The nett effect is thus one of equal change in calcium and Alkalinity, i.e.

Δ Ca = Δ Alk

i.e. Δ Alk - Δ Ca = 0

i.e. Δ (Alk - Ca) = 0

i.e. Δ C_2 = 0

(ii)   From Eq. (21) each part of $Ca(OH)_2$ added decreases the
         Acidity by an equal amount, i.e.

$Ca(OH)_{2\ added}$ = equivalent decrease in Acidity

$$= + \Delta\ C_1$$

Thus, addition of $Ca(OH)_2$ to a saturated water moves the point describing the water vertically in the Modified Caldwell-Lawrence diagram a distance equal to the concentration $Ca(OH)_2$ added expressed as $CaCO_3$.

Example

X ppm $Ca(OH)_2$ as $CaCO_3$ is added to a saturated water whose initial condition plots at Point 1 in Figure 32. From the discussion above: equal changes occur in Alkalinity and calcium concentrations, the final state of the water lies thus on a vertical line through Point 1. Adding X ppm $Ca(OH)_2$ as $CaCO_3$ reduces Acidity by X ppm as $CaCO_3$, i.e. the new condition of the water lies on the horizontal line X ppm as $CaCO_3$ *above* the original Acidity line.

The final saturated condition of the water is given by the values of pH, Alkalinity and Calcium lines through Point 2, Figure 32.

5. <u>Addition of Strong Acid to a Saturated Water in Contact with</u>
   <u>CaCO$_3$</u>

Adding a strong acid to a saturated water in contact with CaCO$_3$
has the following effects:

(i) From Eq. (21) Acidity of the water increases by an
     equal amount, i.e.

H$^+_{added}$ = increase in Acidity

$\qquad\quad$ = - Δ C$_1$

(ii) (a) From Eq. (20) each part of H$^+$ added reduces
           Alkalinity by an equal amount and undersaturated
           conditions develop.

      (b) CaCO$_3$ now dissolves into the water until satura-
           tion is re-established. Each part of CaCO$_3$ dis-
           solving into the water adds a part of each Ca^{++}
           and CO$_3^=$ to the water.

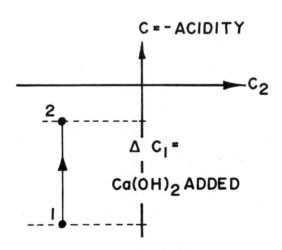

Figure 32. Ca(OH)$_2$ added to a saturated water

Thus from Eq. (40) both Ca^{++} and Alkalinity increase by an equal amount, i.e.

H^+_{added} = decrease in Alkalinity due to (a) + increase
in Alkalinity due to (b)

= decrease in Alkalinity due to (a) + increase
in Ca^{++} due to (b)

= $- \Delta$ Alk + Δ Ca

and Δ denotes an increase.

i.e.

H^+_{added} = Δ(Ca - Alk)

$$= - \Delta C_2 \qquad\qquad (72)$$

Example

Y ppm as $CaCO_3$ of H^+ is added to a water whose initial condition is given by Point 1 in Figure 33. From Eq. (21) Acidity increases by Y ppm as $CaCO_3$, i.e. the horizontal line which is Y ppm *below* the horizontal ordinate passing through Point 1 gives the new Acidity of the water.

Also from Eq. (72):

H^+_{added} = $- \Delta C_2$ = Y ppm as $CaCO_3$

Intersection of the new C_1 and C_2 co-ordinates gives Point 2 in the diagram. The new state of the water is thus given by the values of pH, Alk and Ca lines passing through Point 2, Figure 33.

6. Addition of a Strong Base to a Saturated Water

Adding a strong base (say NaOH) to a saturated water causes the following changes in the water: Alkalinity and pH increase immediately, the water is now oversaturated and $CaCO_3$ precipitates with time, i.e.

(i)   From Eq. (20) each part of OH^- added increases the
      Alkalinity by an equal amount, and for the subsequent
      $CaCO_3$ precipitation, equal parts of Ca^{++} and $CO_3^=$ are
      lost, i.e.

OH^-_{added} = (measured increase in Alkalinity)

           + (Alkalinity lost in $CaCO_3$ precipitated)

      = Δ Alk + CO_3 precipitated

      = Δ Alk - Δ Ca

   where Δ denotes an increase.

i.e.

OH^- added = Δ (Alk - Ca)

        = Δ C_2

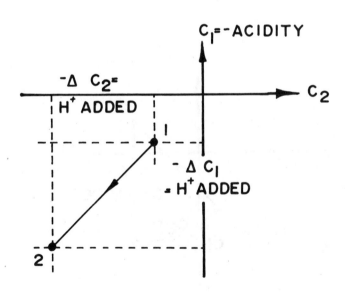

Figure 33. H^+ added to a saturated water in contact with solid
$CaCO_3$

(ii) From Eq. (21) each equivalent of OH$^-$ added decreases
      the Acidity by an equivalent amount, i.e.

$$OH^-_{added} = \Delta C_1$$

Figure 34 shows the effect of adding Z ppm OH$^-$ as CaCO$_3$ to a
saturated water whose initial condition is given by Point 1.

The application of the above derivations to water softening
using the Modified Caldwell-Lawrence diagram is summarized below:

1. The state of a saturated water plots as a point in the conditioning diagram.

2. Adding various chemicals to a saturated water in contact
   with CaCO$_3$ moves the point in the diagram, describing the
   saturated state, a specific distance and direction depending on the type of concentration of chemical added.

3. Choosing C$_1$ and C$_2$ to have equal scales in the diagram

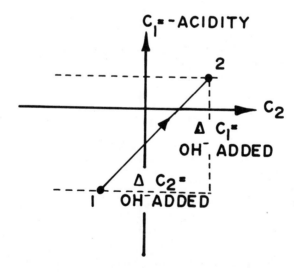

Figure 34. OH$^-$ added to a saturated water

Figure 28, gives the direction which a saturated
equilibrium point in the conditioning diagram moves
due to addition of various chemicals.

Example

Analysis of a water gives: pH 8,4, Alkalinity 55 ppm and Ca 80 ppm
(both as $CaCO_3$). Determine the concentration of $Ca(OH)_2$ required
to adjust the water to a condition where the Ca content is minimum.

The lines representing the measured Alkalinity, calcium and pH
values intersect at Point 1 in Figure 35, i.e. the water is just
saturated with respect to $CaCO_3$. From the direction format diagram, Figure 28, adding $Ca(OH)_2$ moves the point describing the
saturated condition of the water vertically above Point 1 a distance equal to the concentration of $Ca(OH)_2$ added expressed as
$CaCO_3$.

After say 40 ppm $Ca(OH)_2$ (as $CaCO_3$) has been added the final
saturated condition attained is given by Point 2, Figure 35, with
pH 9,2, Alkalinity 18 ppm and Ca 43 ppm (both as $CaCO_3$). The
calcium concentration is now (80 - 43) = 37 ppm less than the
original condition.

Looking along the vertical line through Point 1, one observes
that the calcium value for Points on this line decreases vertically upwards until Point 3 is reached on the zero horizontal
axis with zero Acidity. For Points on this vertical line above
the zero horizontal axis the calcium concentration for saturated
conditions again increases. Thus the final saturated condition
of the water for minimum calcium content is given by Point 3 with
Alkalinity 11 ppm, Ca 36 ppm (both as $CaCO_3$) and pH 9,9. The
required $Ca(OH)_2$ concentration is the vertical distance between
Points 1 and 3, i.e. 55 ppm $Ca(OH)_2$ as $CaCO_3$.

Application of the direction format diagram for calculating
chemical dosage requirements in water softening problems is valid
where adjusting a water from one saturated condition to another

is required. However, the situation arises where a water to be
softened is initially either over- or undersaturated with respect
to $CaCO_3$, i.e. the lines representing the condition of a water do
not intersect in a point in the Modified Caldwell-Lawrence diagram.

If the initial condition of a water is oversaturated, the
point representing its saturated condition is first estimated in
the Modified Caldwell-Lawrence diagram as a mass of $CaCO_3$ will
precipitate from the oversaturated water irrespective of whether

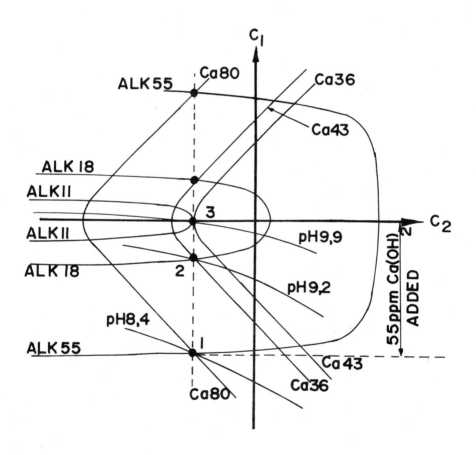

Figure 35. Estimation of $Ca(OH)_2$ dose to soften a water to a
minimum Ca value

$Ca(OH)_2$ is added or not. Thereafter, calculations are carried
out considering this saturated state as being the initial condition of the water.

If the initial condition of a water is undersaturated, dosage
requirements to adjust the water to a saturated state are first
calculated considering the state of the water to be governed by
equilibria between ionic species (i.e. a single aqueous phase
system). Thereafter dosage calculations are carried out in the
Modified Caldwell-Lawrence diagram considering the initial and
final states of the water to be saturated. The total concentration of chemical used to adjust the water to its final condition
is thus the sum of:

(a)   chemical dose required to bring the water to saturation
      considering only single ionic phase equilibria, and

(b)   the concentration of chemical required to adjust the
      water from the saturated condition achieved in (a) to
      some desired saturated condition. For this purpose
      the direction format diagram is used.

Water Softening - Magnesium Removal

The conditioning diagram adequately describes the interdependence
between the maximum calcium content for $CaCO_3$ saturation, the
carbonic species concentrations and pH. However, up to now, the
maximum allowable magnesium content for $Mg(OH)_2$ saturation has
not been interpreted in the diagram. For softening purposes the
maximum concentration of magnesium in water is assumed to be
governed by the solubility product for $Mg(OH)_2$. The solubility
of $MgCO_3$ is so high that it can be considered in the same light
as Na_2CO_3, i.e. as being very soluble. A relationship for the
temperature dependence of this solubility product in the range
$0°C$ to $80°C$ and at unit activity is given in Chapter 3 as:

$pK_{Mg(OH)_2} = 0,0175 \cdot t + 9,97$

where t is in deg. C.

The solubility product for $Mg(OH)_2$ in a water with ionic strength μ and at some temperature between 0°C to 80°C is:

$$(f_D \cdot [Mg^{++}])(f_M \cdot [OH^-])^2 = K_{Mg(OH)_2} \tag{73}$$

where f_D and f_M are the mono- and divalent activity factors respectively, calculated for water with ionic strength μ. [ ] indicate molar concentrations.

Rearranging Eq. (73):

$$[Mg^{++}][OH^-]^2 = \frac{K_{Mg(OH)_2}}{f_D \cdot f_M^2} \quad , \quad \text{or}$$

$$[Mg^{++}][OH^-]^2 = K'_{Mg(OH)_2} \tag{74}$$

where $K'_{Mg(OH)_2}$ is the active solubility product.

Solving for Mg^{++} in Eq. (74):

$$[Mg^{++}] = \frac{K'_{Mg(OH)_2}}{[OH^-]^2}$$

i.e.

$$[Mg^{++}] = \frac{K'_{Mg(OH)_2}}{(K'_w)^2} \cdot [H^+]^2 \tag{75}$$

where K'_w is the ionization constant for water at the temperature in question and ionic strength μ.

From Eq. (38), Chapter 3

$$pH = - \log_{10}(f_M[H^+])$$

Thus, for each pH there is some limiting maximum concentration

of magnesium which cannot be exceeded, or stated another way -
for each magnesium concentration there is some pH which cannot be
exceeded without $Mg(OH)_2$ precipitating.

In each conditioning diagram is a nomogram based on the equations (Eq.(75) and Eq. (38), Chapter 3) relating maximum concentration of magnesium (expressed as ppm $CaCO_3$) with pH for temperature and ionic strength constant.

Effect of $Mg(OH)_2$ Precipitation in the Modified Caldwell-Lawrence Diagram

Precipitation of $Mg(OH)_2$ from a $CaCO_3$ saturated water in contact
with $CaCO_3$ creates the following changes in the condition of the
water:

(i)   Each part of $Mg(OH)_2$ precipitated removes a part of
      each Mg^{++} and OH^- from solution, i.e.

Mg^{++} precipitated = OH^- precipitated

From Eq. (21):

OH^- precipitated = Acidity increase

i.e.

Mg^{++} precipitated = Acidity increase

$$= -\Delta\ C_1 \qquad\qquad (76)$$

where Δ denotes an increase.

(ii) Loss of OH^- from solution in the $Mg(OH)_2$ precipitate
      reduces the pH in the water and causes a concentration
      of $CaCO_3$ to re-dissolve. Thus, for changes in Alka-
      linity two effects are noted:

(a)   For $Mg(OH)_2$ precipitation

Mg^{++} precipitated = OH^- precipitated

and from Eq. (20):

OH^- precipitated = Alkalinity precipitated

i.e.

Mg^{++} precipitated = Alkalinity precipitated

(b) For $CaCO_3$ re-dissolving:

Ca^{++} dissolved = $CO_3^=$ dissolved

= Alkalinity dissolved

The nett effect of (a) and (b) is:

Mg^{++} precipitated = Alkalinity precipitated

+ {Ca^{++} dissolved - Alkalinity dissolved}

= Δ Ca - Δ Alk

where Δ denotes an increase

i.e.

Mg^{++} precipitated = Δ {Ca - Alk}

$$= - \Delta C_2 \tag{77}$$

Thus in the Modified Caldwell-Lawrence diagram, $Mg(OH)_2$ precipitation has a similar effect as adding a strong acid to water in contact with $CaCO_3$, i.e. the point describing saturation moves downwards to the left at 45°C. $Mg(OH)_2$ precipitation continues until its solubility product is again satisfied.

For magnesium removal from water the usual problem encountered in practice is one in which the initial Mg^{++} concentration is known and it is required to calculate chemical dosages to condition a water to some specified final Mg^{++} concentration.

Example

Initial condition of a saturated water is given by Point 1 in Figure 36. Mg^{++} concentration is 60 ppm as $CaCO_3$. Using $Ca(OH)_2$ soften the water to a final Mg^{++} concentration of 5 ppm as $CaCO_3$.

(i) Mg^{++} to be precipitated is $(60 - 5) = 55$ ppm as $CaCO_3$.
     Noting that the initial C_2 ordinate value of the
     water in the diagram is $- 10$ ppm as $CaCO_3$, from Eq.
     (77) the final C_2 ordinate value after $Mg(OH)_2$ pre-
     cipitation is $(- 10 - 55) = - 65$ ppm as $CaCO_3$.

(ii) Referring to a pH-Mg^{++} nomogram, the final required
     pH of the water for saturation with 5 ppm Mg^{++} dis-
     solved is 11,1, i.e. after 55 ppm Mg^{++} has precipitated

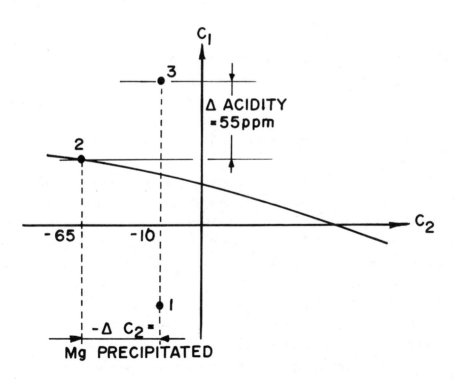

Figure 36. Precipitation of magnesium (as $Mg(OH)_2$) by adding $Ca(OH)_2$

the final condition of the water lies on the pH line
11,1, i.e. the final state of the water is given by
Point 2 in Figure 36.

(iii)   Calculating $Ca(OH)_2$ requirements. From the direction
        format diagram, Figure 28, addition of $Ca(OH)_2$ moves
        Point 1 vertically in the conditioning diagram. Also,
        precipitation of 55 ppm $Mg(OH)_2$ as $CaCO_3$ increases
        the Acidity of the water by an equivalent amount,
        i.e. prior to precipitation of $Mg(OH)_2$ the condition
        of the water is given by Point 3 in Figure 36.

Thus, $Ca(OH)_2$ requirement is given by the difference in ordinate levels between Points 1, and 3, Figure 36.

Example

Water has the following quality: Alkalinity 140 ppm, calcium 240
ppm, magnesium 40 ppm (all as ppm $CaCO_3$) and pH 7,6. The temperature of the water is 20°C and the ionic strength 0,01. Calculate the concentration of $Ca(OH)_2$ and Na_2CO_3 required to adjust the magnesium concentration to 4 ppm and calcium 60 ppm
(both as $CaCO_3$)
    The initial state of the water plots as Point 1 in the modi-
fied Caldwell-Lawrence diagram Figure 37, i.e. the water is
initially saturated with $CaCO_3$. Referring to the direction format diagram, Figure 28, addition of $Ca(OH)_2$ moves Point 1 vertically upwards a distance equal to the concentration of $Ca(OH)_2$
added (as ppm $CaCO_3$).
    The magnesium-pH nomogram in Figure 37 shows at pH 10,8 the
water is just saturated with 40 ppm Mg^{++} (as ppm $CaCO_3$). Increasing the pH above 10,8 causes precipitation of $Mg(OH)_2$.
    Final stipulated magnesium concentration is 4 ppm (as ppm
$CaCO_3$). The magnesium-pH nomogram in Figure 37 shows that at
saturation with pH 11,3 the maximum concentration of magnesium

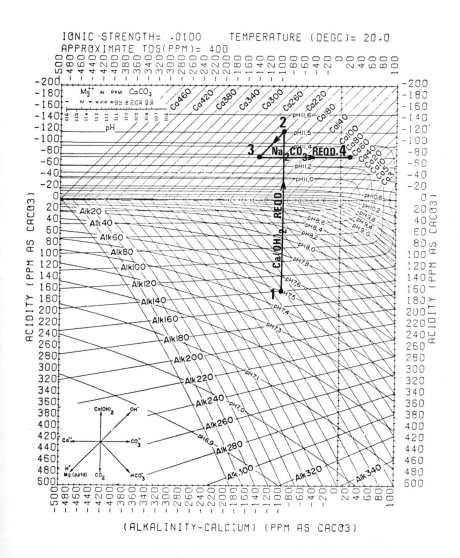

Figure 37. Water softening using $Ca(OH)_2$ and Na_2CO_3

in the water is 4 ppm. Thus, in this problem, after precipitating $(40 - 4) = 36$ ppm $Mg(OH)_2$ (as ppm $CaCO_3$) we require a pH of 11,3.

Ca(OH)$_2$ dosage required to achieve a pH of 11,3 (i.e. 4 ppm magnesium) after precipitation of $Mg(OH)_2$ is calculated as follows:

In the conditioning diagram draw the line representing the pH value at which the water is just saturated with the final desired magnesium concentration, i.e. at pH 11,3 (Figure 37) the water is just saturated with 4 ppm magnesium (as ppm $CaCO_3$).

Initial (Alk-Ca) ordinate value (i.e. C_2 value) in the water from Figure 37 is - 105 (ppm $CaCO_3$).

Precipitation of 36 ppm $Mg(OH)_2$ as $CaCO_3$ causes the final condition of the water to have a C_2 ordinate value of $(- 105 - 36)$ $= - 141$ ppm as $CaCO_3$. Intersection of this C_2 ordinate value with pH line 11,3 gives Point 3 in Figure 37 as the condition of the water after 36 ppm $Mg(OH)_2$ has precipitated.

To calculate Ca(OH)$_2$ required to effect this precipitation, Point 2 in Figure 37 and Figure 38 is established as follows:

(i)   Adding Ca(OH)$_2$ to a saturated water moves the point describing saturation vertically in the diagram, i.e Point 2 lies vertically above Point 1.

(ii) Acidity of Point 2 is 36 ppm as $CaCO_3$ less than Acidity value of Point 3, i.e. Ca(OH)$_2$ required = Acidity Point 1 - Acidity Point 2 = 280 ppm as $CaCO_3$.

Calcium concentration (given by Point 3 in Figure 37) is now 225 ppm (as ppm $CaCO_3$), a final calcium concentration of 60 ppm is specified. Referring to Figure 28, addition of 165 ppm Na_2CO_3 moves Point 3 on Figure 37 horizontally to the right to Point 4. Thus, with time, 175 ppm $CaCO_3$ precipitates and the condition of the water will eventually be Alkalinity 78 ppm, Ca 60 ppm (both as ppm $CaCO_3$), pH 11,3. Carbon dioxide would normally be added to the water at this stage finally to adjust pH, Alkalinity and

and Ca to desired values.

EQUILIBRIUM BETWEEN CO_2 IN THE AIR AND CARBONIC SPECIES IN SOLUTION

CO_2 exchange between water and the atmosphere takes place until the CO_2 partial pressures in the two phases are equal, i.e. *at equilibrium between air and water the dissolved CO_2 concentration is fixed.*

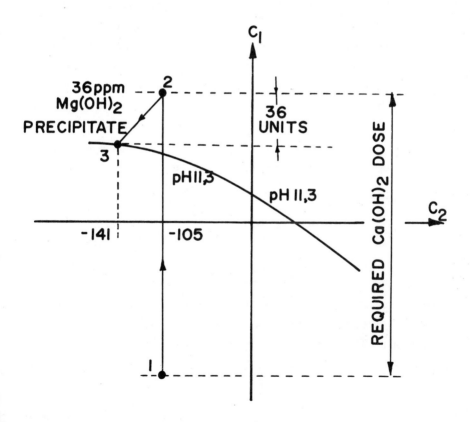

Figure 38. Magnesium precipitation using $Ca(OH)_2$

In the process of advancement to this equilibrium the pH in the water changes and there is a redistribution of the dissolved carbonic concentrations, i.e. a change in the dissolved CO_2 concentration occurs and more CO_2 is exchanged with the air. The pH at which equilibrium is established depends on the Alkalinity of the water. Exchange of CO_2 between air and water does not change the Alkalinity only the Acidity and pH provided no $CaCO_3$ precipitation occurs.

The rate of CO_2 transfer depends on the difference in partial pressure of CO_2 across the air-water interface. With mixing, surface renewal of CO_2 over- or undersaturated water at the interface increases the transfer rate and thus decreases the time required for a body of water to achieve CO_2 equilibrium with the air. In tranquil bodies of water CO_2 equilibrium between air and water may be achieved only at the water-air interface and chemical changes in the body of water will be controlled by the rate of diffusion of the CO_2.

For equilibrium between dissolved and atmospheric CO_2 at a particular partial pressure of CO_2 (P_{CO_2}) the concentration of dissolved CO_2 is defined by Henry's Law as:

$$[CO_2] = K'_{CO_2} \cdot P_{CO_2} \tag{78}$$

K'_{CO_2} is Henry's Law Constant which is temperature dependent

P_{CO_2} is the partial pressure of CO_2 in the atmosphere.

More convenient is to have the lefthand side of Eq. (78) in terms of $H_2CO_3^*$, the sum of dissolved CO_2 and H_2CO_3 in the water. From Eq. (8), Chapter 3:

$$\frac{[H_2CO_3^*]}{[CO_2]} = K_c$$

where K_c is a constant which is independent of temperature in the range $0°C$ to $50°C$.

i.e.

$$[H_2CO_3^*] = [CO_2].K_c \tag{79}$$

Substituting Eq. (78) into Eq. (79):

$$[H_2CO_3^*] = K'_{CO_2} \cdot P_{CO_2} \cdot K_c$$

i.e.

$$[H_2CO_3^*] = K_{CO_2} \cdot P_{CO_2} \tag{80}$$

where the constant K_{CO_2} is temperature dependent.

From data supplied by Hamer (1961), a relationship between K_{CO_2} and temperature is developed and shown in Figure 39. From this plot, the data can be approximated by two linear functions:

For the range $0°C$ to $35°C$

$$pK_{CO_2} = 1,12 + 0,0138.t$$

and for the range $35°C$ to $80°C$

$$pK_{CO_2} = 1,36 + 0,0069.t$$

Expressing the concentration of $H_2CO_3^*$ in Eq. (80) as ppm $CaCO_3$:

$$[H_2CO_3^*] = H_2CO_3^*/10^5 \qquad \text{i.e.}$$

$$H_2CO_3^*/10^5 = K_{CO_2} \cdot P_{CO_2} \qquad \text{i.e.}$$

$$H_2CO_3^* = K_{CO_2} \cdot P_{CO_2} \cdot 10^5 \qquad \text{i.e.}$$

$$H_2CO_3^* = K_{cCO_2} \cdot P_{CO_2} \tag{81}$$

For CO_2 equilibrium between the atmosphere and water, Alkalinity and pH are directly related as follows (Weber and Stumm, 1963):

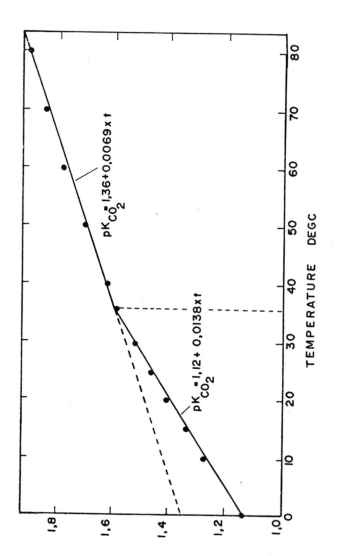

Figure 39.   Effect of temperature on CO_2 equilibrium between air and water

From Eq. (10):

$$HCO_3^- = H_2CO_3^* \cdot K'_{c1}/H^+ \tag{82}$$

and from Eq. (11):

$$CO_3^= = HCO_3^- \cdot K'_{c2}/H^+ \tag{83}$$

Substituting Eq. (82) into Eq. (83):

$$CO_3^= = H_2CO_3^* \cdot K'_{c1} \cdot K'_{c2}/H^2 \tag{84}$$

Substituting Eq. (81) into each of Eqs. (82 and 84) gives:

$$HCO_3^- = K'_{cCO_2} \cdot P_{CO_2} \cdot K'_{c1}/H^+ \quad \text{and} \tag{85}$$

$$CO_3^= = K'_{cCO_2} \cdot P_{CO_2} \cdot K'_{c1} \cdot K'_{c2}/H^{+2} \tag{86}$$

Substituting Eqs. (85 and 86) into Alkalinity equation, Eq. (7), gives Alkalinity as a function of only the hydrogen ion concentration for constant temperature, ionic strength and partial pressure of CO_2 in the air:

$$Alk = \frac{K_{cCO_2} \cdot P_{CO_2} \cdot K'_{c1}}{H^+} (1 + K'_{c2}/H^+) + K'_{cW}/H^+ - H^+ \tag{87}$$

i.e.

$$Alk = f(H^+) = f_1(pH)$$

Thus, from Eq. (87) Alkalinity is directly related to pH for a water at equilibrium with CO_2 in the air.

Interpretation in Modified Caldwell-Lawrence Diagram

Each conditioning diagram gives a nomogram of pH against Alkalinity for CO_2 equilibrium between air and water. Plotting the nomogram values gives a two phase equilibrium line, Line A in Figure 40. A water whose condition (i.e. whose intersection of pH and Alkalinity defined as the ionic equilibrium point) plots

on Line A is at equilibrium with CO_2 in the air, e.g. Alkalinity
96 ppm and pH 8,5; Alkalinity 75 ppm and pH 8,4. If, however,
ionic equilibrium plots either above or below Line A, CO_2 exchange
occurs between the water and air.

Example (in which no $CaCO_3$ is precipitated)

Analysis of a water gives Alkalinity 65 ppm, calcium 45 ppm (both
as $CaCO_3$) and pH 8,7. What is the condition of the water after
CO_2 equilibrium with the air is attained?

Initial condition plots above CO_2 equilibrium line A at Point
1, Figure 40. The water is thus not in equilibrium with the CO_2
in the atmosphere but saturated with respect to $CaCO_3$. Vertical
ordinate value of Point 1 gives the Acidity of the water as 62
ppm as $CaCO_3$.

As CO_2 exchange occurs between the air and water the following
effects are noted:

(i) From Eq. (20), loss or gain of CO_2 by a water does
      not change the Alkalinity, i.e.

Final Alkalinity = initial Alkalinity

$$= 65 \text{ ppm as } CaCO_3$$

The pH and Acidity in the water, however, do change.

The final pH in the water after CO_2 equilibrium is
attained with the air is given by the intersection
of Alkalinity 65 ppm line and the two phase equilibrium line, line A, Point 2 in Figure 40, i.e. pH 8,34.

(ii) From Eq. (21):

$$\text{Acidity change} = CO_{2 \text{ added}} + HCO_{3 \text{ added}}^- + H_{\text{added}}^+ - OH_{\text{added}}^-$$

Thus, as a water attains CO_2 equilibrium with the air
the change in Acidity gives the concentration of CO_2
exchanged. Final Acidity is given by the vertical

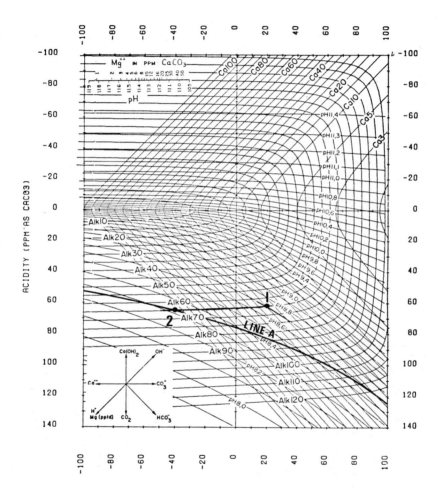

IONIC STRENGTH= .0100   TEMPERATURE (DEGC)= 15.0

APPROXIMATE TDS(PPM)= 400

(ALKALINITY-CALCIUM) (PPM AS CACO3)

Figure 40.   Water becomes undersaturated with $CaCO_3$ as CO_2 equilibrium is attained with the air

ordinate value of Point 2, Figure 40, i.e. 66 ppm
as $CaCO_3$.

Acidity change = final Acidity - initial Acidity

$$= 66 - 62 = + 4 \text{ ppm as } CaCO_3$$

i.e. an increase in Acidity means CO_2 is absorbed
by the water.

CO_2 absorbed = Δ Acidity = 4 ppm as $CaCO_3$

(iii) The value of Ca line intersecting Point 2 in Figure
40 (i.e. Ca 100 ppm) is the required calcium concentration for $CaCO_3$ saturation. The water, however,
contains only 45 ppm of calcium, i.e. the water has
become undersaturated with respect to $CaCO_3$.

In general, a water whose ionic equilibrium point plots above
the CO_2 equilibrium line absorbs CO_2 from the atmosphere and becomes undersaturated with respect to $CaCO_3$, as CO_2 equilibrium
is attained with the air.

Example (in which $CaCO_3$ is precipitated)

Analysis of a water gives Alkalinity 100 ppm, calcium 90 ppm
(both as $CaCO_3$) and pH 8,2. What is the final condition of the
water after CO_2 equilibrium is attained with the air?

Initial state of the water plots below CO_2 equilibrium line A
at Point 1 in Figure 41. The water is thus not in equilibrium
with the CO_2 in the air but is saturated with respect to $CaCO_3$.

As CO_2 exchange occurs between air and water the following
effects are noted:

(i) From Eq. (20) loss or gain of CO_2 does not change the
Alkalinity of the water. Intersection of lines Alkalinity 100 and Line A (i.e. Point 2, Figure 41) gives
final pH in the water, i.e. pH 8,53.

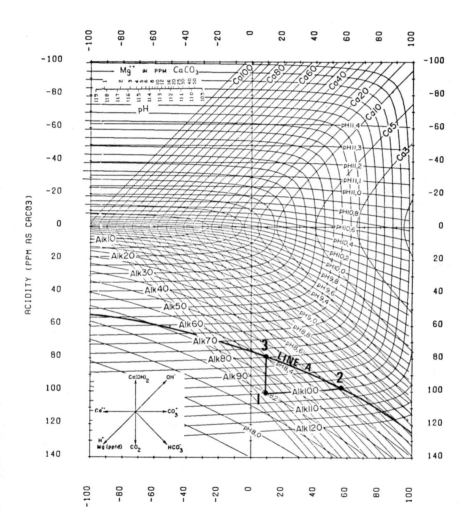

Figure 41. Water precipitates $CaCO_3$ as CO_2 equilibrium is attained with the air

(ii)   As a water attains CO_2 equilibrium with the air,
       Eq. (21), the change in Acidity gives the concen-
       tration of CO_2 exchanged.

Initial Acidity = 102 ppm as $CaCO_3$

Final Acidity   = 97 ppm as $CaCO_3$ i.e.

Acidity change = (97 - 102) = - 5 ppm as $CaCO_3$

  i.e. 5 ppm of CO_2 (as $CaCO_3$) is expelled from the
  water to the air.

(iii) The value of the calcium line intersecting Point 2
       in Figure 41 (i.e. Ca 43 ppm) gives the required
       calcium concentration for $CaCO_3$ saturation. The
       water, however, contains 90 ppm of Ca and is there-
       fore oversaturated causing $CaCO_3$ precipitation.
       Alkalinity decreases and the water is no longer in
       equilibrium with CO_2 in the air. Eventually the
       water attains a state of three phase equilibrium -
       between dissolved carbonic species, $CaCO_3$ and CO_2
       in the air, i.e. Point 3, occurring on the two
       phase equilibrium line and vertically above Point
       2 in Figure 41.

 Final state of the water is:

Alkalinity 78 ppm, Ca 68 ppm (both as $CaCO_3$) and pH 8,42.

 $CaCO_3$ precipitated is given by the change in either Alkalinity or calcium concentration

= (100 - 78) = (90 - 68) = 22 ppm $CaCO_3$

 CO_2 expelled from the water equals the decrease in Acidity,
i.e.

CO_2 expelled = (101 - 79) = 22 ppm as $CaCO_3$

 In general, water whose initial condition plots below two

phase equilibrium line precipitates $CaCO_3$ as equilibrium with CO_2 in the air is attained.

Example

An undersaturated water becomes oversaturated with respect to $CaCO_3$.

Analysis of a water gives: Alkalinity 95 ppm, Ca 70 ppm and pH 8,2. What final state will the water attain with time?

Intersection of Alkalinity and pH gives initial ionic equilibrium point, Point 1 in Figure 42. Calcium concentration required for $CaCO_3$ saturation is 95 ppm as $CaCO_3$ and the water is thus initially undersaturated.

When CO_2 is exchanged with the atmosphere, Alkalinity does not change and the new ionic equilibrium point is given by intersection of lines Alkalinity 95 ppm and Line A (Point 2, Figure 42) with new pH 8,5. Calcium concentration required for $CaCO_3$ saturation is now 35 ppm as $CaCO_3$, i.e. the water is oversaturated with respect to $CaCO_3$ and precipitation occurs. Further CO_2 transfer and $CaCO_3$ precipitation occurs until three phase equilibrium is attained, i.e. Point 4 on Line A vertically above Point 3 in Figure 42. Final state of the water is:

Alkalinity 85 ppm, Ca 60 ppm (both as $CaCO_3$) and pH 8,45

$CaCO_3$ precipitated is given by the change in either Alkalinity or calcium caoncentration, i.e.

$CaCO_3$ precipitated = $(70 - 60)$ = $(95 - 85)$ = 10 ppm $CaCO_3$

CO_2 expelled from the water is given by the decrease in Acidity, i.e.

CO_2 expelled = $(97 - 83)$ = 14 ppm as $CaCO_3$

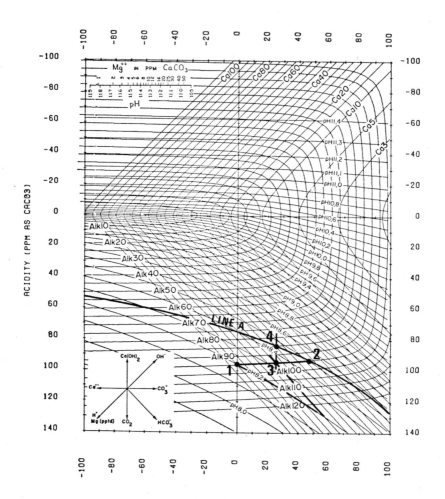

Figure 42. A CaCO₃ undersaturated water becomes oversaturated
with CO₂ exchange between air and water

References

Arbatsky,J.W.	1940	"Zeichnerische Ermittlung der Enthärt-ungsverhältnisse von Wassern", Gas-u, Wässerface, <u>83</u>, 90, 116.
Caldwell,D.H. and Lawrence,W.B.	1951	"Water Softening and Conditioning Problems", Ind.Eng.Chem., <u>45</u>, 3, 535
Hamer,P, Jackson, J. & Thurston,E.G.	1961	*Industrial Water Treatment Practice*, Butterworth (Co. (Publishers) Ltd., London.
Stumm,W. and Weber,W.J.	1963	"Mechanism of Hydrogen Ion Buffering in Natural Waters, Jour.A.W.W.A., <u>55</u>, 1553.

Chapter 6

ALKALINITY-CALCIUM DEFICIENT WATERS

A characteristic frequently encountered in waters derived from the Table Mountain sandstone areas in South Africa is that the waters contain extremely low concentrations of calcium and Alkalinity, and have a low pH. An example of such a water was reported at Worcester, South Africa, by Lawson and Snyders (1962). In this water the concentrations of calcium and Alkalinity were each about 2 ppm as $CaCO_3$ and the pH about 6. At Wemmershoek, Steenbras, Constantia Nek and Kloof Nek, all reservoirs serving Cape Town, the water, after floculation and filtration, have a calcium concentration of about 9 ppm, Alkalinity of about 2 ppm (both expressed as $CaCO_3$) and a pH of about 6.

If pH, calcium and Alkalinity of such waters are plotted in a conditioning diagram, the intersection of pH and Alkalinity lines occurs at some point to the left, off the diagram, and there is no intersection of the calcium with the Alkalinity line. The latter observation (non-intersection) differentiates these waters from the normal undersaturated waters where calcium-Alkalinity intersection occurs. Waters which do not show intersection have certain features requiring special consideration when theoretically calculating dosages for conditioning. For this reason they are given a special designation, *Alkalinity-calcium deficient waters*.

The group of Alkalinity-calcium deficient waters can be subdivided into two groups:

(a) <u>Waters with relatively high pH (greater than about 8,5)</u>

It is unlikely that such a condition will be found in natural waters, for a water governed by the carbonic system and having a relatively high pH will plot in a conditioning diagram about the 3-phase equilibrium line - thus to reach equilibrium CO_2 is absorbed from the

air. The absorbed CO_2 serves as an acid source causing
a reduction in pH and creating a water which is able
to dissolve more carbonates and hence increases the
Alkalinity. It is thus unlikely that a water open to
the atmosphere, or other sources of CO_2, will have both
low Alkalinity and high pH.

(b) Waters with low pH

Typical of these waters are those derived from the Table
Mountain sandstone areas of South Africa. Such waters
cannot be treated to *calcium carbonate saturation* by addition of only either calcium ions or hydroxide ions
without increasing either of these concentrations to
inordinately high concentrations. To bring such waters
to a pH range in which they are potable requires the addition of carbonic species, hence they are designated
as '*carbonic species deficient water*'.

Carbonic species deficient waters have a low Alkalinity
and calcium concentration, low pH and are undersaturated
with respect to $CaCO_3$ – these qualities make the water
extremely aggresive to both metal and concrete pipes and
undesirably soft. As such they are unsuitable for domestic purposes and it is essential that they be stabilized to give a well-buffered slightly oversaturated
water.

Stabilization of Carbonic Species Deficient Water Using the
Modified Caldwell-Lawrence Diagram
Calculations for estimating the mass of chemicals required for
stabilizing carbonic species deficient water cannot be carried
out in the Modified Caldwell-Lawrence Diagram in the straightforward manner described in Chapter 5 for 'normal' water. The
reason for this is that for these waters the initial Acidity

cannot be estimated in the diagram as the intersection of lines
representing the measured Alkalinity and pH values occurs to the
left off the diagram. An estimation of this initial Acidity
value, vital for all dosage calculations, is obtained from the
Alkalinity-Acidity-pH equilibrium diagram developed in Chapter 5
(see Figure 1). Referring to Figure 1, the intersection point
of lines representing the measured initial Alkalinity and pH
values gives the initial Acidity value.

Once the initial Acidity value is known, calculations for estimating chemical dosages required to stabilize the water to any
desired Alkalinity, pH and calcium values are determined using
the modified Caldwell-Lawrence diagram.

Stabilization of Carbonic Species Deficient Water

Calculations for stabilizing carbonic species deficient water
are carried out in two steps:

1. Concentration of dosing chemicals required to adjust
    the parameters Alkalinity, pH and calcium to speci-
    fied values are estimated considering the water to
    be governed only by equilibrium between dissolved
    carbonic species (i.e. aqueous phase equilibria).

2. A quantitative measure of the degree of over- or
    undersaturation of the final state of the water is
    estimated considering equilibrium between two phases
    - solid $CaCO_3$ and dissolved carbonic species.

Example (in which $Ca(OH)_2$ and CO_2 are the prescribed stabilizing
chemicals)

Analysis of a water gives Alkalinity 2 ppm, Ca^{++} 6 ppm (both as
$CaCO_3$) and pH 6,1. Determine the concentrations of $Ca(OH)_2$ and
CO_2 required to adjust the water to Alkalinity 30 ppm and pH
9,1.

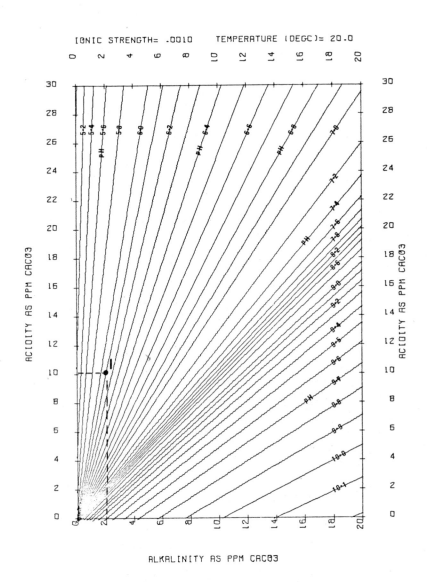

Figure 1. Intersection of lines for Alkalinity 2 ppm and pH 6,1 gives Acidity 10 ppm

1. (a) Initial Acidity in the water is estimated in the
        Alkalinity-Acidity-pH equilibrium diagram, Figure
        1, i.e.

        Initial Acidity = 10 ppm as $CaCO_3$

   (b) Concentration of $Ca(OH)_2$ required to adjust Alka-
        linity to the prescribed calue is found as follows:

        Alkalinity required = Prescribed Alk - Initial Alk

                            = (30 - 2) = 28 ppm as $CaCO_3$

        From Table 2, Chapter 5, and Eq. (20) Chapter 5:

        Each part of $Ca(OH)_2$ added increases both Alkali-
        nity and calcium concentration by one part, i.e.

        $Ca(OH)_2$ required = Alkalinity required

                            = 28 ppm as $CaCO_3$

        New Alkalinity = 2 + 28 = 30 ppm as $CaCO_3$

        New Ca^{++} = 6 + 28 = 34 ppm as $CaCO_3$

   (c) pH in the water after adjusting Alkalinity to
        30 ppm by adding 28 ppm (as $CaCO_3$) is found as
        follows:

        Initial Acidity, from (a) above = 10 ppm as $CaCO_3$

        From Eq. (21), Chapter 5, adding 28 ppm $Ca(OH)_2$
        as $CaCO_3$ decreases Acidity by an equal amount, i.e.

        Acidity change   = - 28 ppm as $CaCO_3$

        New Acidity   = - 28 + 10 = - 18 ppm as $CaCO_3$

        The value of the pH line through the intersection
        point of Alkalinity 30 ppm and Acidity - 18 lines,
        Point 1, Figure 2, defines the new pH, i.e. pH
        10,75.

(d) pH is adjusted to the value pH 9,1 by adding CO_2 gas to the water. Adding CO_2 to the water has the following effects:

(i) From Eq. (21), Chapter 5, Acidity increases by an amount equal to CO_2 added.

(ii) From Eq. (20), Chapter 5, Alkalinity does not change (provided $CaCO_3$ is not precipitated).

The final ionic equilibrium state after CO_2 addition is given by the intersection of the lines representing pH 9,1 and Alkalinity 30 ppm, Point 2, Figure 2. Acidity ordinate value of Point 2 defines the final Acidity in the water, i.e. Acidity 27 ppm as $CaCO_3$.

$$CO_2 \text{ required} = \text{Final Acidity} - \text{Acidity after } Ca(OH)_2 \text{ addition}$$

$$= 27 - (-18) = 45 \text{ ppm as } CaCO_3$$

Because the final specified state may not necessarily be just saturated with respect to $CaCO_3$ one expects the desired final state to be defined by three parameters, say Alkalinity, pH and Ca^{++}. However, in this problem only two parameters are specified in defining the final desired stabilized condition - Alkalinity and pH. This observation is explained by noting that where $Ca(OH)_2$ and CO_2 are the conditioning chemicals Alkalinity and calcium concentration cannot be varied independently of each other.

Adding either or both of these two chemicals causes the Alkalinity and Ca^{++} to change by equal amounts whether or not $CaCO_3$ precipitation occurs, thus prescribing a desired Alkalinity or calcium value automatically defines the value of the other.

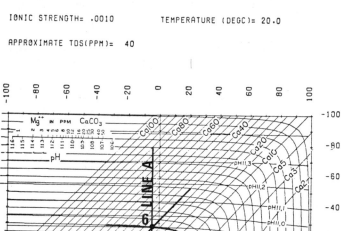

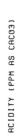

Figure 2. Stabilization of Cape water using Ca(OH)$_2$ and CO$_2$

2. Estimation of the saturation state of the water. The
    value of the calcium line through Point 2, Figure 2,
    gives the theoretically required calcium concentration
    for $CaCO_3$ saturation, i.e. Ca^{++} 23 ppm as $CaCO_3$. Com-
    paring this theoretically required calcium value with
    the actual calcium value (34 ppm as $CaCO_3$) one deduces
    that the water is oversaturated and $CaCO_3$ precipitates
    with time. The final saturated state which the water
    ultimately achieves is found in Figure 2 from the values
    of Alkalinity, Ca^{++} and pH lines through Point 4, i.e.
    Alkalinity 27 ppm, Ca^{++} 31 ppm (both as $CaCO_3$) and pH
    9,0.

    The mass of $CaCO_3$ which precipitates as the water achieves
    saturated equilibrium is given by the difference between
    the saturated and oversaturated value of either Alkalinity
    or Ca^{++}, i.e.

    $CaCO_3$ precipitated $= (30 - 27) = (34 - 31)$

    $\qquad\qquad\qquad\qquad = 3$ ppm

In water stabilization problems the parameter (Alk-Ca) can
often be used to simplify graphical solution procedures in the
modified Caldwell-Lawrence diagram. In the above example where
water is stabilized using $Ca(OH)_2$ and CO_2, Alkalinity and Ca^{++}
values always change by equal amounts and hence the parameter
(Alk-Ca) has a constant value, i.e.

    Initial Alkalinity and Ca^{++} values of the water are
    2 ppm and 6 ppm (as $CaCO_3$) respectively.

    (Alk-Ca) $= (2 - 6) = - 4$ ppm as $CaCO_3$ and plots as
    Line A in Figure 2.

    Final Alkalinity and Ca^{++} values for the stabilized
    water are 30 ppm and 34 ppm (as $CaCO_3$) respectively.

(Alk-Ca) = (30 - 34) = - 4 ppm as $CaCO_3$ which is
the same value as the initial condition.

Furthermore, one observes that the lines representing Alkalinity 30 ppm and calcium 34 ppm intersect at Point 3 in Figure 2 which lies on Line A.

In general, where $Ca(OH)_2$ and CO_2 are the stabilizing chemicals the parameter (Alk-Ca) has a constant value and provided that the lines representing Alkalinity and calcium values intersect, their point of intersection always occurs on the vertical ordinate line representing their (Alk-Ca) value, i.e. Points 3, 4, 5 and 6 in Figure 2. The vertical ordinate value of this intersection point on the (Alk-Ca) line depends on the concentrations of chemicals added.

For each calcium line in the conditioning diagram there is a lower limiting Alkalinity curve which is just tangential to the calcium line on the horizontal axis. For example, in Figure 2, the Alkalinity line which is just tangential to the calcium line, Ca = 17 ppm, is Alkalinity = 13 ppm (both as $CaCO_3$), i.e. any Alkalinity line with a value less than 13 ppm will not intersect the Ca 17 ppm line, and, irrespective of the pH, such a water cannot be saturated with respect to $CaCO_3$ without increasing Alkalinity or calcium or both. In the example above $Ca(OH)_2$ is added to the water to adjust Alkalinity and calcium concentrations to desired values; (Alk-Ca) has a constant value of - 4 ppm as $CaCO_3$ and provided that the lines representing Alkalinity and calcium values intersect, this intersection occurs on the line (Alk-Ca) = - 4 ppm. If only 11 ppm of $Ca(OH)_2$ (as $CaCO_3$) were added to the water:

Alkalinity = 2 + 11 = 13 ppm as $CaCO_3$

Calcium = 6 + 11 = 17 ppm as $CaCO_3$

(Alk-Ca) = (13 - 17) = - 4 ppm as $CaCO_3$

Plotting the Alkalinity and calcium lines representing these
values the two lines intersect tangentially on Line A at Point 5
in Figure 2. If less than 11 ppm of $Ca(OH)_2$ (as $CaCO_3$) is added
no intersection of the lines is obtained. Thus the values of
Alkalinity and calcium lines at the point of intersection of the
specified (Alk-Ca) value and the horizontal axis gives the absolute minimum Alkalinity and Ca^{++} values required such that the pH
in the water can be adjusted (by adding CO_2) to obtain a saturated water.

Experimental Verification of Dosage Prediction

Treatment of Cape Town Water

After flocculation and filtration, water from Steenbras and Kloof
Nek dams has approximately the following condition: Alkalinity
2 ppm, calcium 9 ppm (both as $CaCO_3$) and pH 6,1. The water is
stabilized in two stages, initial treatment with lime to a desired Alkalinity and calcium concentration, and final pH adjustment to a desired value using CO_2. The approximate final condition of the water is Alkalinity 28 ppm, calcium 35 ppm (both as
$CaCO_3$) and pH 9,0. The water is oversaturated by approximately
+ 0,2 on the Langelier Index. Dosage requirements for this or
any other final state can easily be determined in a conditioning
diagram. However, if only $Ca(OH)_2$ and CO_2 are used as dosing
chemicals, the final calcium concentration is not independent of
Alkalinity, so that either Alkalinity or calcium must be specified. Also the pH depends on the degree of oversaturation required.

In Cape Town, the Alkalinity and a degree of oversaturation
(on the Langelier Index) are specified. These values were probably originally developed taking into account that the final pH
had to be about 9. In other words, the condition of the treated
Cape Town water, as it is specified at present, was probably

arrived at by experiment with due regard to having a pH within a
desired range.

Calculations for Estimating Dosage Requirements

$Ca(OH)_2$ and CO_2 requirements to adjust Cape Town water to the
condition Alkalinity 27 ppm (as $CaCO_3$) and Saturation Index + 0,2
are calculated using equilibria diagrams (for a mean summer water
temperature of 20°C) as follows:

(a) Estimation of the Acidity of the raw water. Referring
    to Alkalinity-Acidity-pH equilibrium diagram, Figure 1,
    Acidity in the raw water is 10 ppm as $CaCO_3$.

(b) $Ca(OH)_2$ requirements. Cape Town desires an Alkalinity
    of 27 ppm (as $CaCO_3$), hence (27 - 2) = 25 ppm $Ca(OH)_2$
    (as $CaCO_3$) must be added to the water. Thus new Ca^{++}
    value = (9 + 25) = 34 ppm as $CaCO_3$. Intersection of
    Ca 34 and Alkalinity 27 lines occurs at Point 1, Figure 3.

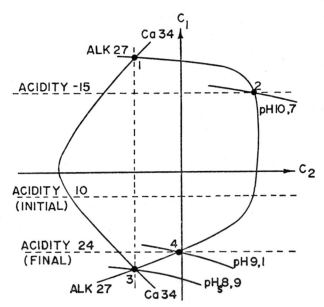

Figure 3. Stabilization of Cape water by adjusting pH to 9,1
and S.I. to + 0,2

(c) pH after adding 25 ppm $Ca(OH)_2$ as $CaCO_3$. Change in
Acidity after $Ca(OH)_2$ dosing = - 25 ppm as $CaCO_3$.
New Acidity = (+10 - 25) = - 15 ppm as $CaCO_3$. Intersection of the lines for Acidity - 15 ppm and
Alkalinity 27 ppm occurs at Point 2 in Figure 3. The
value of the pH line through Point 2 defines the new
pH as 10,7.

(d) Carbon dioxide dosage requirements. Adding CO_2 to the
water: Acidity increases by an amount equal to CO_2
added (as $CaCO_3$), Alkalinity remains unchanged. Now
the concentration of CO_2 to be added depends on the
required Langelier Index. If CO_2 is added until the
water is just saturated, pH is given by the value of
the pH line through the intersection of Alkalinity 27
ppm and calcium 34 ppm lines, i.e. pH line through
Point 3 vertically below Point 1 in Figure 3. This
gives a pH_s of 8,9. As a Langelier Saturation Index
of + 0,2 is specified, the final pH should be pH = 8,9
+ 0,2 = 9,1. The point of intersection of pH 9,1 with
Alkalinity 27 ppm plots at Point 4, Figure3. Acidity
ordinate value of Point 4 gives the final Acidity in
the water, i.e. Final Acidity = 24 ppm as $CaCO_3$. CO_2
requirement is given by the difference in Acidity between Points 2 and 4, i.e.

CO_2 required = Final Acidity - Acidity after adding $Ca(OH)_2$

$$= 24 - (- 15) = 39 \text{ ppm as } CaCO_3$$

Total dosage requirements are therefore:

$Ca(OH)_2$ = 25 ppm as $CaCO_3$

CO_2 = 39 ppm as $CaCO_3$

One can now compare the dosages and final pH values of theoretical and experimental data.

	Theoretical	Experimental
$Ca(OH)_2$ (ppm as $CaCO_3$)	25	25
CO_2 (ppm as $CaCO_3$)	39	*
pH	9,1	9,1
Langelier Index	+ 0,2	+ 0,2

* CO_2 is supplied from a gas burner and hence concentration is
  not known.

Perhaps the most important point is that by a simple graphical
construction in the conditioning diagram it is possible to simu-
late dosages and final water condition closely. Consequently
many dosage combinations can be rapidly evaluated.

It is evident, following the dosage addition above, that the
final condition as described by calcium-Alkalinity intersection
point must lie on the vertical line through Points 1 and 3,
Figure 3. Thus if less $Ca(OH)_2$ is added, Point 1 moves downwards and Point 3 upwards so that the final saturated pH value
increases.

A disadvantage of Cape Town's treated water is the extremely
low buffering capacity which is supplied by the carbonic species
in the pH range 8 to 9 (see Figure 7, Chapter 3). This low buffering capacity can give rise to corrosion problems as briefly
discussed in Chapter 4 (see page 130). Referring to the buffer
capacity diagram for the carbonic system one sees that by increasing the pH to about 10, the total buffer capacity of the
water can be increased many-fold.

A scheme to adjust the raw Cape Town water (after flocculation
and filtration) to a final state where the pH is about 10 using
$Ca(OH)_2$ and CO_2 is as follows:

Supposing a treated water with a pH of 10,1 is specified together with a Langelier Saturation Index of + 0,2, then

Langelier's theoretical saturated pH, pH_s, is 10,1 + 0,2 = 10,3.
(The state of the water plots above the zero Acidity line and as
shown in Chapter 5, page 216, the S.I. reverses and in this case
is given by S.I. = pH_s - pH). (Alk-Ca) = 2 - 9 = - 7 ppm as $CaCO_3$.

The specified pH_s 10,3 line is drawn in the diagram and intersects the vertical ordinate line (Alk-Ca) = - 7 at Point 1,
Figure 4.

The final calcium and Alkalinity values are now fixed by the
respective values of their lines that pass through Point 1,
Figure 4, i.e. Alkalinity 13 ppm and calcium 20 ppm (both as
$CaCO_3$).

$Ca(OH)_2$ Dosage

Required $Ca(OH)_2$ dosage to increase Alkalinity and calcium concentrations from 2 ppm and 9 ppm respectively to 13 ppm and 29
ppm respectively is 11 ppm (all as $CaCO_3$).

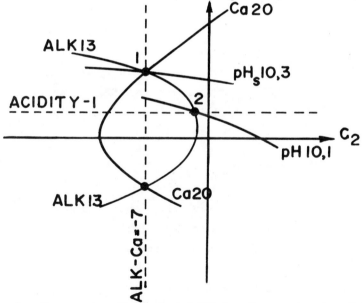

Figure 4. Proposed method of stabilizing Cape water to pH 10,1
and S.I. + 0,2

pH in the water after adding 11 ppm $Ca(OH)_2$ is found as follows:

Acidity of raw water (from Figure 1) = 10 ppm as $CaCO_3$.
Acidity change on adding 11 ppm $Ca(OH)_2$ = - 11 ppm as $CaCO_3$, i.e.
New Acidity = 10 - 11 = - 1 ppm as $CaCO_3$.

The value of the pH line through the intersection of Acidity - 1 ppm and Alkalinity 13 ppm lines, Point 2, Figure 4, gives the value of the new pH, i.e. pH 10,1.

pH 10,1 is the pH value prescribed for the stabilized water and thus in this case no CO_2 is required to adjust the pH.

Examining the final condition of the water:

(a) the $Ca(OH)_2$ addition is about 50 per cent of that required at present and described previously. i.e. 11 ppm (as $CaCO_3$) compared with 28 ppm (as $CaCO_3$).

(b) the CO_2 addition is nil; 39 ppm (as $CaCO_3$) is required at present.

(c) the buffering index of the proposed water is about 5 times greater than that for the water treated at present, i.e. $3,5*10^{-4}$ moles/pH compared with $7,2*10^{-5}$ moles/pH. This increase in buffer capacity should have marked beneficial effects on corrosion reduction.

Treatment of Worcester Water

In a pilot plant study on conditioning the water of Worcester, South Africa, Lawson and Snyders (1962) showed that a carbonic species deficient water could be stabilized by dosing with a strong acid and powdered calcium carbonate and finally adjusting it with lime.

The state of the water entering the pilot plant was Alkalinity 2 ppm, calcium 2 ppm (both as $CaCO_3$) and pH 6,0. Treating the water with 90 ppm $CaCO_3$ and 60 ppm H_2SO_4 (as $CaCO_3$) and finally

adjusting it with 21 ppm Ca(OH)$_2$ yielded an oversaturated water
with the following condition:

Alkalinity 59 ppm, calcium 111 ppm (both as CaCO$_3$), pH 8,6 and
total hardness 115 ppm as CaCO$_3$.

It is of interest to check whether the same final quality of
water can be attained theoretically using a conditioning diagram
for the chemical dosing employed by Lawson and Snyders on this
water.

To use the conditioning diagram, the temperature of the water
and the ionic strength must be estimated, since Lawson and Snyders
do not report these parameters and in the absence of this information:

(a) a temperature of 20°C is assumed (this is the average
     annual water temperature for the draw-off from the
     nearby Wemmershoek impoundment);

(b) ionic strength of the water is estimated from an
     empirical equation proposed by Langelier (1936):

     $\mu = 4H - A$

     where H = hardness in moles/ℓ; and A = Alkalinity
     in equivalents/ℓ.

     For hardness of 0,00115 moles/ℓ and Alkalinity
     0,0006 equivalents/ℓ, ionic strength μ is calcu-
     lated 0,004.

Chemical Dosing

Chemical treatment of this water was carried out in two stages.
In the first stage H$_2$SO$_4$ and powdered CaCO$_3$ were added to the
water. The effect of these two chemicals is to increase the Alkalinity and calcium concentrations in the water. However, the
water remains undersaturated with respect to CaCO$_3$. In the
second stage Ca(OH)$_2$ is added to stabilize the water to an oversaturated condition.

The condition of the water after each of these stages can be theoretically predicted using the conditioning diagram as follows:

<u>1st Stage</u>:   60 ppm H_2SO_4 (as $CaCO_3$) and 90 ppm $CaCO_3$ are added to the water.

Assuming the water remains undersaturated with respect to $CaCO_3$ the new condition of the water is found by considering the state of the water to be governed only by equilibria between dissolved carbonic species, i.e. a single aqueous phase system

(a) . Initial Acidity of the raw water cannot be estimated in the modified Caldwell-Lawrence diagram as intersection of lines representing pH and Alkalinity occurs to the left off the diagram. However, initial Acidity can be estimated in the Alkalinity-Acidity-pH equilibrium diagram, Figure 1:

Initial Acidity = 11 ppm as $CaCO_3$.

(b)   Adding 60 ppm H_2SO_4 (as $CaCO_3$) to the water increases Acidity and decreases Alkalinity each by 60 ppm as $CaCO_3$ i.e.

Alkalinity change = - 60 ppm as $CaCO_3$
Acidity change   = + 60 ppm as $CaCO_3$

(c)   Dissolving 90 ppm of $CaCO_3$ into this water increases Alkalinity and Ca^{++} concentrations each by 90 ppm (as $CaCO_3$); Acidity remains unchanged, i.e.

Alkalinity change = + 90 ppm as $CaCO_3$
Acidity change   = 0
Ca^{++} change   = + 90 ppm as $CaCO_3$

The overall changes in the parameters Alkalinity, Acidity and Ca^{++} resulting from acid addition and $CaCO_3$ dissolution is given by the sum of the effects in (b) and (c) above, i.e.

Net Alkalinity change = - 60 + 90 = + 30 ppm as $CaCO_3$

Net Acidity change = + 60 + 0 = 60 ppm as $CaCO_3$

Net Ca^{++} change = 0 + 90 = 90 ppm as $CaCO_3$

The new Alkalinity, Acidity and Ca values in the water after the first stage of treatment are estimated by adding the changes in these parameters due to dosing and the initial values of the parameters, i.e.

Alkalinity = Initial Alkalinity + Alkalinity change

= 2 + 30 = 32 ppm as $CaCO_3$

Acidity = Initial Acidity + Acidity change

= 11 + 60 = 71 ppm as $CaCO_3$

Ca^{++} = Initial Ca^{++} + Ca^{++} change

= 2 + 90 = 92 ppm as $CaCO_3$

The new pH in the water is defined in the modified Caldwell-Lawrence diagram by the value of the pH line through the intersection of Acidity 71 ppm and Alkalinity 32 ppm lines. This intersection point occurs to the left off the diagram. However, pH can be estimated in the Alkalinity-Acidity-pH equilibrium diagram, Figure 5, i.e. pH = 6,55. Table 1 lists and compares the measured and theoretical conditions in the water after the first stage of treatment.

Table 1.

Condition of Worcester water after adding 60 ppm H_2SO_4 (as $CaCO_3$) and 90 ppm $CaCO_3$

	Measured	Theoretical
Alkalinity (ppm as $CaCO_3$)	36	32
Ca^{++} (ppm $CaCO_3$)	87	92
pH	6,4	6,55

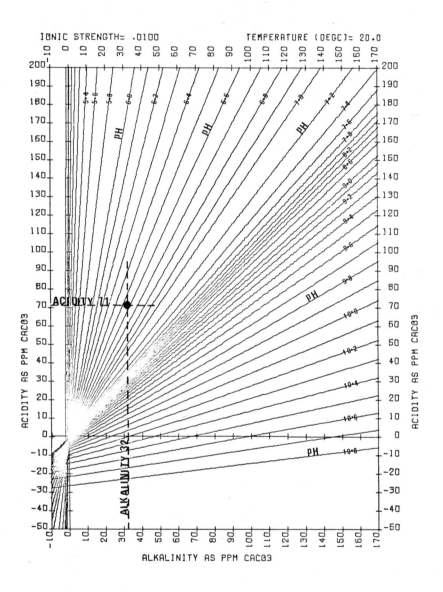

Figure 5.   pH of Worcester water after CaCO$_3$ and acid addition
is pH 6,55

Plotting the lines representing pH, Alkalinity and Ca values
of the water in a conditioning diagram shows the water to be
undersaturated with respect to $CaCO_3$. Final stabilization is
now attained using $Ca(OH)_2$.

2nd Stage: 21 ppm $Ca(OH)_2$ (as $CaCO_3$) is added to the water.

Assuming no $CaCO_3$ precipitation occurs the new condition of
the water is found theoretically as follows:

Adding 21 ppm $Ca(OH)_2$ (as $CaCO_3$) to the water increases Alkalinity and Ca^{++} concentrations each by 21 ppm as $CaCO_3$ and decreases Acidity by 21 ppm (as $CaCO_3$), i.e.

Alkalinity change = + 21 ppm (as $CaCO_3$)
Acidity change   = - 21 ppm (as $CaCO_3$)
Ca^{++} change   = + 21 ppm (as $CaCO_3$)

The final stabilized values of the parameters Alkalinity,
Acidity and Ca^{++} are found by adding the changes in these parameters due to $Ca(OH)_2$ addition and the values of these parameters
after the 1st stage of treatment, i.e.

Final Alkalinity = Alkalinity after 1st stage + Alkalinity
                   change due to $Ca(OH)_2$ addition

                 = 32 + 21 = 53 ppm as $CaCO_3$

Final Acidity   = Acidity after 1st stage + Acidity change
                   due to $Ca(OH)_2$ addition

                 = 71 - 21 = 50 ppm as $CaCO_3$

Final Ca^{++} = Ca^{++} after 1st stage + Ca^{++} change due
                   to $Ca(OH)_2$ addition

                 = 92 + 21 = 113 ppm as $CaCO_3$

The final pH in the water is found using the modified Caldwell-
Lawrence diagram as follows:

Intersection of Alkalinity 53 ppm and Acidity 50 ppm lines
occurs at Point 1, Figure 6. The value of the pH line through
Point 1 defines the final pH in the water, i.e. pH 8,75.

Plotting the lines representing Alkalinity, pH and Ca values
of the water in a conditioning diagram indicates that the water
is oversaturated with respect to $CaCO_3$ (see Figure 6). A quantitative estimate of the degree of oversaturation is found as
follows: Final saturated state which the water ultimately
achieves is given by the values of Alkalinity, calcium and pH
lines through Point 2, Figure 6, i.e. Alkalinity 49 ppm, calcium
109 ppm (both as $CaCO_3$) and pH 8,2.

The mass of $CaCO_3$ which ultimately precipitates from the water
is the difference between the oversaturated and saturated values
of either Alkalinity or calcium (expressed as ppm $CaCO_3$), i.e.

$$CaCO_3 \text{ oversaturation} = (53 - 49) = (113 - 109)$$

$$= 4 \text{ ppm}$$

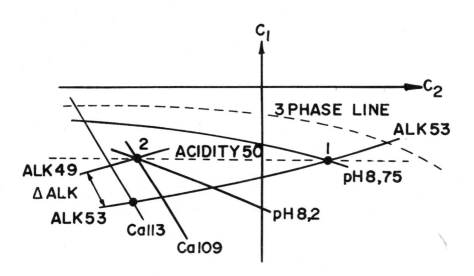

Figure 6. Final stabilized Worcester water after lime addition

Theoretical and measured values for the final state of the water are listed and compared in Table 2.

Table 2.

	Measured	Theoretical
Alkalinity (ppm as $CaCO_3$)	59	53
Calcium (ppm as $CaCO_3$)	111	113
pH	8,9	8,75

Experimental calcium value is slightly less than the theoretical value; this is probably due to some $CaCO_3$ having precipitated and settled out of solution. Experimental pH value is slightly less than the theoretical value; this could be due to CO_2 exchange with the atmosphere. The final state of the water plots below the three phase equilibrium line and there is a tendency for CO_2 to be expelled from the water. In the pH range 8 to 9 the carbonic species offers very little pH buffer capacity and a small concentration of CO_2 expelled can cause relatively large pH changes. For this water, loss of only about 1 ppm CO_2 (as $CaCO_3$) causes a pH increase of about 0,15. Experimental Alkalinity is about 10 per cent greater than theoretical value. If some precipitation had taken place (as indicated by Ca^{++}), experimental Alkalinity value should be less than the theoretical value of 53 ppm. The most likely explanation for this difference is that the Alkalinity measurement was inaccurate.

Repeating the graphical procedures above in condition diagrams for temperatures 15°C and 25°C indicate no significant improvement in correlation between measured data and theoretical values.

Verification in the Laboratory

The single, two and three phase conditioning models presented in the Modified Caldwell-Lawrence diagram were each tested experimentally in the laboratory. All tests were carried out on water at 25°C with an ionic strength of approximately 0,01 M.

Experimental procedure

Six tests were conducted. The first four tests consisted of two parts (a) single aqueous phase tests and (b) two phase tests. The final two tests each also included a third part - a three phase equilibrium test.

Each water sample used in these tests was prepared as follows:

(i)   Distilled water was fed through a commercial deionizer which also removed CO_2 from the water.

(ii) NaCl was dried at 110°C, cooled in a desiccator and weighed to four decimal places into a porcelain dish.

(iii) The NaCl was washed into a 2 litre flask which was then filled to the mark with CO_2 free de-ionized water. The ionic strength of this solution was then 0,01 M.

The following solutions were prepared as chemical additives to be used in the respective tests:

(i) Na_2CO_3 - 0,04 M (4003 ppm as $CaCO_3$)

(ii) $NaHCO_3$ - 0,08 M (4003 ppm as $CaCO_3$)

(iii) NaOH   - 0,08 M (3955 ppm as $CaCO_3$)

(iv) $CaCl_2$ - 0,10 M (10125 ppm as $CaCO_3$)

The Na_2CO_3 salt was first dried at 110°C and cooled in a desiccator. The $NaHCO_3$ was dried in a desiccator only. The

requisite mass of this salt was weighed and made up to 1 litre of
solution with CO_2 free de-ionized water. The analar Na_2CO_3 and
$NaHCO_3$ solutions are standard solutions and thus do not need to
be standardized against any other chemical. To prepare the calcium solution, $CaCl_2.2H_2O$ was used and neither heated nor dried
prior to dilution to 1 litre with CO_2 free de-ionized water. This
solution was standardized against E.D.T.A. which, in turn, had
been standardized against a standard calcium solution.

500 mℓ of the water sample was transferred into a reaction
flask which was immersed in a water bath held at a constant temperature of 25°C. The water at this stage contained no Alkalinity
or Acidity.

The experimental apparatus used is listed at the end of this
chapter.

(a) Single Phase Test:

A selected mass in solution of chemical (i), (ii) or
(iii) was added to the reactor vessel and the initial
Alkalinity and Acidity values calculated according to
Eqs. (20 and 21), Chapter 5. The pH of the water was
then observed. This was compared with the pH value
predicted in the Modified Caldwell-Lawrence diagram
for the calculated Acidity and Alkalinity values.
Differing masses of chemical solutions (i), (ii) or
(iii) were then added, the pH being noted after each
addition. The calculated Alkalinity and Acidity
values were corrected for changes in volume due to
the volume added of chemical solution. The water in
the flask was stirred continuously using a magnetic
stirrer.

Experimental results are compared with predicted
values for the single phase tests in Tables 3 to 8,
each table referring respectively to Figures 7 to 12.

(b) Two Phase (precipitation) Test:

To each of the above completed single phase tests a selected mass of $CaCl_2$ was now added to create a condition supersaturated with respect to $CaCO_3$.

Powdered $CaCO_3$ seed (Calcite) was added to ensure that precipitation of $CaCO_3$ from solution was fairly rapid. The mixture was stirred for four hours to allow saturated equilibrium to be attained and pH was then noted. This value was compared with the saturated equilibrium pH value predicted in the diagram.

Differing masses of chemicals (i), (ii), (iii) or (iv) were then added, the pH being noted approximately four hours after each addition. Use of the direction format diagram in the Modified Caldwell-Lawrence diagram yielded the predicted value of pH attained at saturation. The pH values observed are compared with the predicted pH values in Tables 9 to 12 and Figures 13 to 16 respectively.

(c) Three Phase Tests:

The final two tests of the two phase experiments reported above were terminated at a high pH (approximately 10,5). Dried, filtered air was then sparged through the water in the reaction flask to obtain equilibrium between CO_2 in the air and dissolved CO_2 in the water; i.e. three phase equilibrium. The pH was noted and compared with that predicted in the Modified Caldwell-Lawrence diagram using Line A,Figure 17. These results are reported in Tables 13 and 14.

Experimental Results

Comparison of observed with theoretically predicted results for
single, two and three phase experiments are set out in Tables 1
to 14.

In general the experimental pH values observed and plotted
in the diagrams were always slightly below the theoretically
calculated pH values. In order to investigate this observation,
initial Alkalinity and initial observed pH were assumed correct.
This yielded an Acidity value which was on the average 3 ppm (as
$CaCO_3$) above the theoretical initial Acidity.

This Acidity could have been introduced inadvertently into
the sample during the experimental preparations. In order to
test this hypothesis, water which was initially free of CO_2 was
introduced into the reactor vessel in the same fashion as adopted
in the test procedure. The pH of this water was monitored before and after transfer. A decrease in pH (from pH 7,0 to pH
5,7) occurred in the distilled water. Calculation indicated
that approximately 3 ppm (as $CaCO_3$) of CO_2 had been gained by
the water.

This test justified the use of an "observed" initial Acidity
value, obtained from the intersection in the diagram of the
initial Alkalinity line and the initial observed pH line. When
the "observed" initial Acidity value was used in the calculations for chemical dosage, observed pH values and theoretically
calculated pH values agreed to within 0,02 pH units. The diagram also predicted the pH value the water would achieve after
recarbonation of the water with air, to within 0,02 pH units.

TABLE 3.

Test 1. (stabilization, see Fig. 7)

Initial Volume 500,00 ml

Chemical added	Volume or mass added	Alk.	Acidity*			pH*		
			A	B	C	A	B	C
NaHCO$_3$	0,0504g	60,1	60,1	61,5		8,08	7,96	
NaHCO$_3$	3,84 ml	90,1	90,1	91,6	91,9	8,18	8,10	8,10
Na$_2$CO$_3$	3,88 ml	120,0	89,4	92,3	91,2	9,38	9,33	9,35
NaOH	3,94 ml	149,6	58,3	63,7	60,0	10,00	9,94	9,98

* Acidity (A) and pH (A): Acidity and pH predicted from the mass of chemicals added

Acidity (B): Acidity predicted in the diagram from Alkalinity and observed pH

pH (B): Observed pH

Acidity (C) and pH (C): Predicted in the diagram assuming *initial* Acidity (B) to be correct

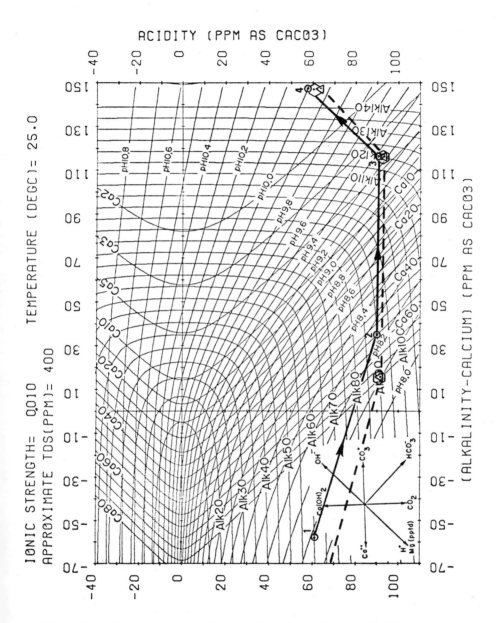

Figure 7. Single Aqueous Phase, Test 1: ○ theoretically predicted values; △ observed values; ⬡ predicted values based on
initial Acidity calculated from initial Alkalinity and observed
initial pH.

TABLE 4.

Test 2. (stabilization, see Fig. 8)

Initial Volume 500,00 ml

Chemical added	Volume or mass added	Alk.	Acidity*			pH*		
			A	B	C	A	B	C
NaHCO$_3$	0,0420g	50,0	50,0			8,16	7,83	
NaOH	1,28 ml	60,0	39,8	44,8	44,6	9,49	0,34	9,34
Na$_2$CO$_3$	2,64 ml	80,7	39,6	44,7	44,6	9,78	9,69	9,69
NaHCO$_3$	2,58 ml	100,7	59,8	64,9	64,8	9,65	9,56	9,56

* Acidity (A) and pH (A): Acidity and pH predicted from the mass of chemicals added

Acidity (B): Acidity predicted in the diagram from Alkalinity and observed pH

pH (B): Observed pH

Acidity (C) and pH (C): Predicted in the diagram assuming *initial* Acidity (B) to be correct

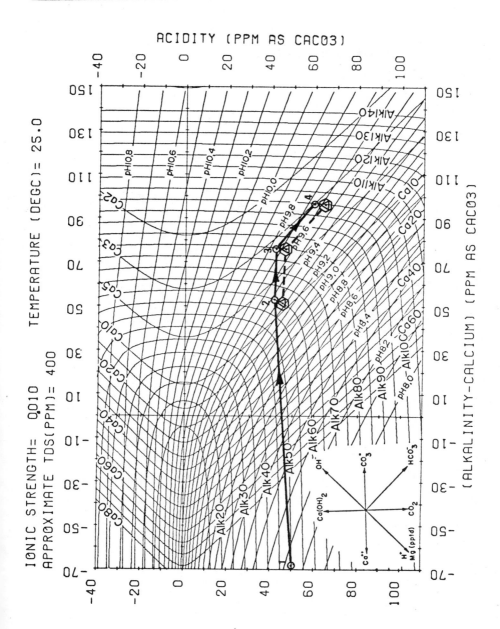

Figure 8. Single Aqueous Phase, Test 2: ◯ theoretically predicted values; △ observed values; ⬡ predicted values based on
initial Acidity calculated from initial Alkalinity and observed
initial pH.

TABLE 5.

Test 3. (stabilization, see Fig. 9)

Initial Volume 500,00 ml

Chemical added	Volume or mass added	Alk.	Acidity*			pH*		
			A	B	C	A	B	C
NaOH	1,27 ml							
Na_2CO_3	1,26 ml	20,0	-10,0	-7,0		10,33	10,26	
Na_2CO_3	3,82 ml	50,1	-9,9	-6,8	-6,9	10,48	10,41	10,42
NaOH	1,28 ml	59,9	-19,9	-15,8	-16,9	10,65	10,60	10,62
$NaHCO_3$	9,18 ml	130,0	51,6	56,8	54,6	9,97	9,92	9,94

* Acidity (A) and pH (A): Acidity and pH predicted from the
                                    mass of chemicals added

  Acidity (B): Acidity predicted in the diagram from Alkali-
                                    nity and observed pH

  pH (B): Observed pH

  Acidity (C) and pH (C): Predicted in the diagram assuming
                                    initial Acidity (B) to be correct

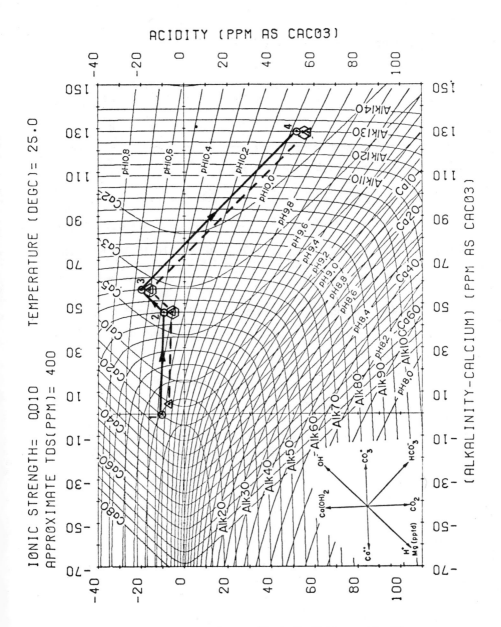

Figure 9. Single Aqueous Phase, Test 3: ◯ theoretically predicted values; △ observed values; ⬡ predicted values based on initial Acidity calculated from initial Alkalinity and observed initial pH.

TABLE 6.

Test 4. (stabilization, see Fig. 10)

Initial Volume 500,00 ml

Chemical added	Volume or mass added	Alk.	Acidity*			pH*		
			A	B	C	A	B	C
Na_2CO_3	3,78 ml	30,0	0	3,2		10,15	10,06	
$NaHCO_3$	7,72 ml	90,0	60,4	62,2	63,6	9,52	9,48	9,46
NaOH	2,66 ml	109,9	39,8	43,7	43,0	10,01	9,96	9,97
Na_2CO_3	3,98 ml	139,8	39,5	45,3	42,7	10,15	10,09	10,12

* Acidity (A) and pH (A): Acidity and pH predicted from the mass of chemicals added

Acidity (B): Acidity predicted in the diagram from Alkalinity and observed pH

pH (B): Observed pH

Acidity (C) and pH (C): Predicted in the diagram assuming *initial* Acidity (B) to be correct

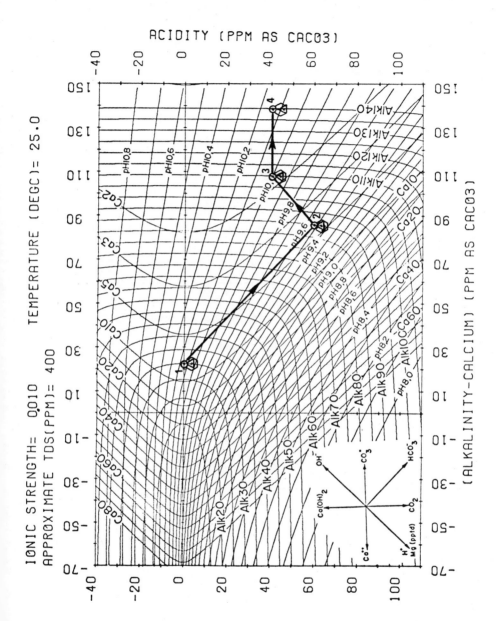

Figure 10. Single Aqueous Phase, Test 4: ◯ theoretically predicted values; △ observed values; ⬡ predicted values based on
initial Acidity calculated from initial Alkalinity and observed
initial pH.

TABLE 7.

Test 5. (stabilization, see Fig. 11)

Initial Volume 500,00 mℓ

Chemical added	Volume or mass added	Alk.	Acidity*			pH*		
			A	B	C	A	B	C
Na$_2$CO$_3$	3,78 mℓ	30,0	0,0	3,15		10,15	10,06	
NaHCO$_3$	7,72 mℓ	90,0	60,4	62,01	63,6	9,53	9,48	9,46
NaOH	2,66 mℓ	109,9	39,8	42,4	42,9	10,01	9,98	9,97
Na$_2$CO$_3$	3,98 mℓ	139,8	39,5	42,6	42,6	10,15	10,12	10,12

* Acidity (A) and pH (A): Acidity and pH predicted from the
                                            mass of chemicals added

   Acidity (B): Acidity predicted in the diagram from Alkali-
                                            nity and observed pH

   pH (B): Observed pH

   Acidity (C) and pH (C): Predicted in the diagram assuming
                                            initial Acidity (B) to be correct

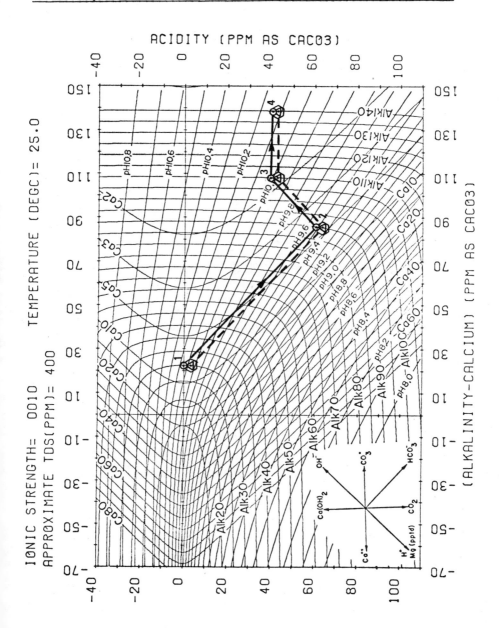

Figure 11.   Single Aqueous Phase, Test 5: ◯ theoretically predicted values; △ observed values; ⬡ predicted values based on
initial Acidity calculated from initial Alkalinity and observed
initial pH.

TABLE 8.

Test 6. (stabilization, see Fig. 12)

Initial Volume 500,00 ml

Chemical added	Volume or mass added	Alk.	Acidity*			pH*		
			A	B	C	A	B	C
Na_2CO_3	3,78 ml	30,0	0,0	2,32		10,15	10,09	
$NaHCO_3$	7,72 ml	90,0	60,4	62,1	62,75	9,53	9,49	9,48
NaOH	2,66 ml	109,9	39,8	42,1	42,1	10,01	9,98	9,98
Na_2CO_3	3,98 ml	139,8	39,5	41,8	41,8	10,15	10,12	10,12

* Acidity (A) and pH (A): Acidity and pH predicted from the mass of chemicals added

Acidity (B): Acidity predicted in the diagram from Alkalinity and observed pH

pH (B): Observed pH

Acidity (C) and pH (C): Predicted in the diagram assuming *initial* Acidity (B) to be correct

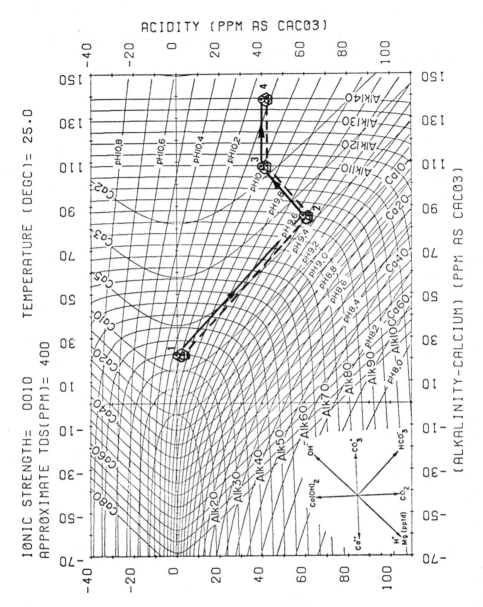

Figure 12.   Single Aqueous Phase, Test 6: ⊖ theoretically predicted values; △ observed values; ⬡ predicted values based on initial Acidity calculated from initial Alkalinity and observed initial pH.

TABLE 9.

Test 3. (cont.) (softening, two phase equilibrium, see Fig. 13)

Initial Volume 516,81 ml

Chemical added	Volume or mass added	(Alk-Ca)	Acidity*			pH*		
			A	B	C	A	B	C
(seed added)		130,00	51,6	56,8	54,6	9,98	9,92	9,94
$CaCl_2$	5,15 ml	28,8	51,1	55,3	54,1	8,79	8,71	8,72
NaOH	4,00 ml	58,4	20,9	25,4	23,9	9,97	9,88	9,90
Na_2CO_2	4,00 ml	88,2	20,7	27,1	23,7	10,16	10,07	10,12
$NaHCO_3$	4,00 ml	117,6	50,5	55,3	53,52	9,93	9,87	9,90

*   Acidity (A) and pH (A): Acidity and pH predicted from the mass of chemicals added

   Acidity (B): Acidity predicted in the diagram from Alkalinity and observed pH

   pH (B): Observed pH

   Acidity (C) and pH (C): Predicted in the diagram assuming *initial* Acidity (B) to be correct

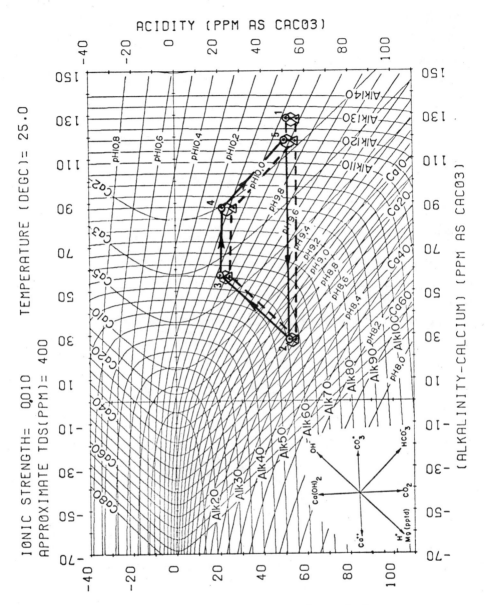

Figure 13.   Two Phase Equilibrium, Test 3 (cont.): ○ theoretically predicted values; △ observed values; ⬡ predicted values
based on initial Acidity calculated from initial Alkalinity
and observed initial pH.

TABLE 10.

Test 4.(cont.) (softening, two phase equilibrium, see Fig. 14)

Initial Volume 518,14 ml

Chemical added	Volume or mass added	(Alk-Ca)	Acidity*			pH*		
			A	B	C	A	B	C
(seed added)		139,8	39,5	45,3	42,7	10,15	10,09	10,12
CaCl$_2$	5,15 ml	38,8	39,1	42,5	42,3	9,27	9,17	9,17
CaCl$_2$	3,96 ml	-37,8	38,8	45,0	42,0	8,45	8,36	8,40
NaHCO$_3$	5,00 ml	-0,5	76,0	83,1	79,2	8,18	8,10	8,15

* Acidity (A) and pH (A): Acidity and pH predicted from the mass of chemicals added

Acidity (B): Acidity predicted in the diagram from Alkalinity and observed pH

pH (B): Observed pH

Acidity (C) and pH (C): Predicted in the diagram assuming *initial* Acidity (B) to be correct

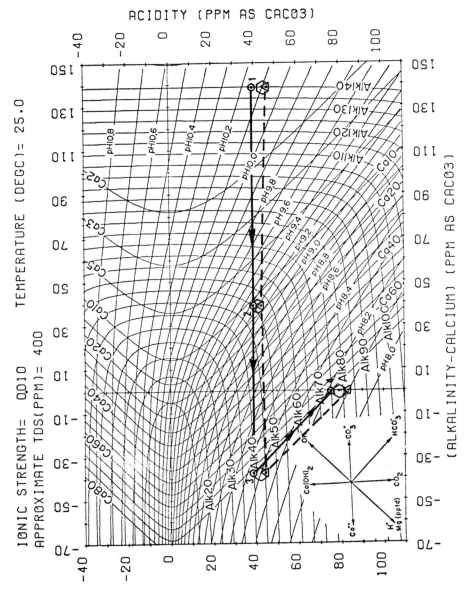

Figure 14.   Two Phase Equilibrium, Test 4 (cont.): ○ theoretically predicted values; △ observed values; ⬡ predicted values
based on initial Acidity calculated from initial Alkalinity
and observed initial pH.

TABLE 11.

Test 5.(cont.) (softening, two phase equilibrium, see Fig.15)

Initial Volume 518,14 ml

Chemical added	Volume or mass added	(Alk–Ca)	Acidity*			pH*		
			A	B	C	A	B	C
(seed added)		139,80	39,5	42,6	42,6	10,15	10,12	10,12
$CaCl_2$	5,15 ml	38,9	39,1	42,3	42,2	9,27	9,18	9,18
$CaCl_2$	3,96 ml	-37,8	38,8	43,2	41,9	8,45	8,38	8,40
$NaHCO_3$	5,00 ml	-0,5	76,0	80,0	79,2	8,18	8,14	8,14
NaOH	11,00 ml	78,7	-5,2	-2,4	-2,0	10,50	10,47	10,46

*   Acidity (A) and pH (A): Acidity and pH predicted from the
                             mass of chemicals added

    Acidity (B): Acidity predicted in the diagram from Alkali-
                  nity and observed pH

    pH (B): Observed pH

    Acidity (C) and pH (C): Predicted in the diagram assuming
                             initial Acidity (B) to be correct

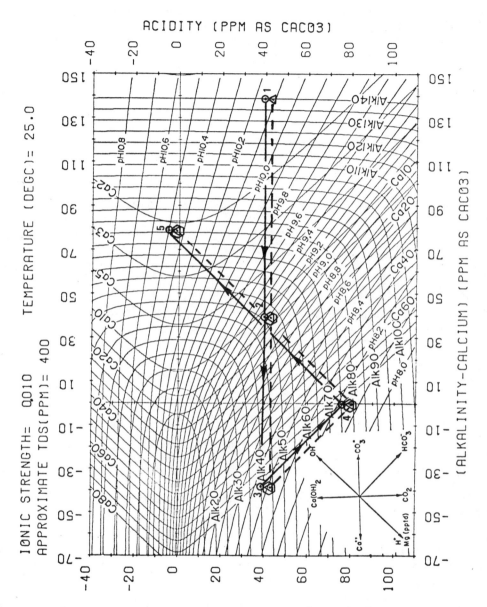

Figure 15. Two Phase Equilibrium, Test 5 (cont.): ○ theoretically predicted values; △ observed values; ◯ predicted values
based on initial Acidity calculated from initial Alkalinity
and observed initial pH.

TABLE 12.

Test 6.(cont.) (softening, two phase equilibrium, see Fig. 16)

Initial Volume 518,14 ml

Chemical added	Volume or mass added	(Alk-Ca)	Acidity*			pH*		
			A	B	C	A	B	C
(seed added)		139,8	39,5	41,8	41,8	10,15	10,12	10,12
CaCl$_2$	5,15 ml	39,0	39,1	42,0	41,4	9,26	9,19	9,19
NaOH	5,3 ml	77,5	-0,8	2,3	1,6	10,43	10,39	10,40
CaCl$_2$	3,7 ml	6,4	-0,8	1,6	1,6	10,05	9,96	9,96
NaOH	4,1 ml							
Na$_2$CO$_3$	2,05 ml	51,5	-30,7	-28,4	-28,4	10,78	10,74	10,75

*   Acidity (A) and pH (A): Acidity and pH predicted from the
                             mass of chemicals added

    Acidity (B): Acidity predicted in the diagram from Alkali-
                  nity and observed pH

pH (B): Observed pH

Acidity (C) and pH (C): Predicted in the diagram assuming
                         initial Acidity (B) to be correct

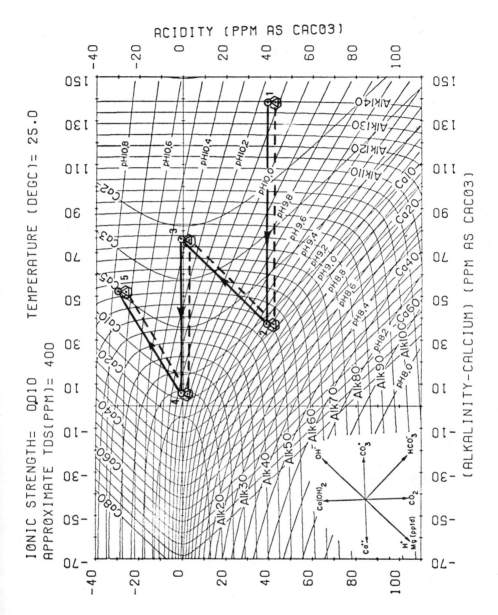

Figure 16. Two Phase Equilibrium, Test 6 (cont.): O theoretically predicted values; △ observed values; ◇ predicted values
based on initial Acidity calculated from initial Alkalinity
and observed initial pH.

TABLE 13.

Test 5.(cont.) (three phase equilibrium, see Fig. 17)

	(Alk-Ca)	Acidity	pH
air	78,7	-2,4	10,47
bubbled			
through	78,7	102,9	8,58 → (8,60 predicted)

TABLE 14.

Test 6.(cont.) (three phase equilibrium, see Fig. 17)

	(Alk-Ca)	Acidity	pH
air	51,5	-28,4	10,75
bubbled			
through	51,5	85,0	8,46 → (8,48 predicted)

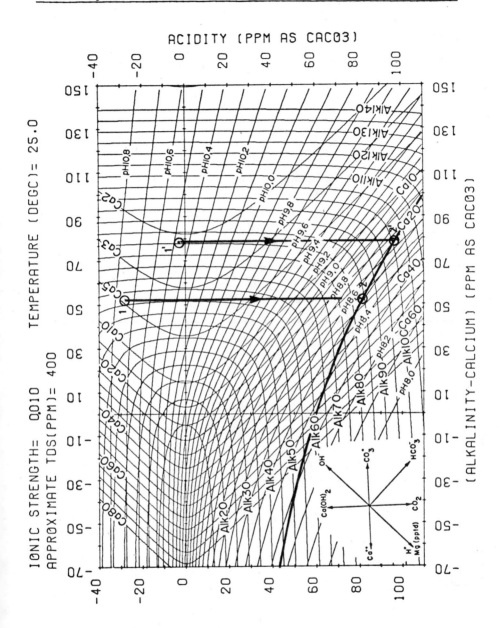

Figure 17.   Three Phase Equilibrium, Tests 5 and 6 (cont.).

Apparatus

Reactor vessel	:	Pyrex glass reaction flask; 3 necked; capacity 500 mℓ
Burette	:	Metrohm piston burette graduated to 0,02 mℓ
pH meter	:	pHM G4 Research pH Meter; Radiometer Copenhagen. Electronic digital display to 3 decimal places
Water bath	:	Perspex container; water kept at constant temperature 25°C
Magnetic stirrer	:	Inert magnet in reaction flask to obtain good mixing
Seed	:	$CaCO_3$. MERCK "proanalys"
pH probe	:	Combined glass electrode, Radiometer Copenhagen. GK 2311 C

Note: The buffers used to standardize the pH probe were primary standard buffers and only rated to an accuracy of $\pm$ 0,01 pH units. As such, no pH readings observed can be guaranteed to an accuracy greater than $\pm$ 0,01.

References

Langlier, W.F.   1936   "The Analytical Control of Anti-corrosion Water Treatment", J.A.W.W.A., 34, 1667.

Lawson, S.P. and
Snyders, R.   1962   "Lime Acid Reaction for Water Stabilization", J.A.W.W.A., 54, 176.

Chapter 7

TREATMENT OF WATER FOR HOT WATER SYSTEMS

Industrial and domestic water is often required at temperatures
well above ambient, say up to 90°C. (Waters in boilers are not
considered here). This requirement creates the following chemical problem: to what state should a water be treated at ambient
temperature so that at an elevated temperature it is neither undersaturated nor too oversaturated with respect to $CaCO_3$? This
type of problem can only be solved using equilibria chemistry.
Analytical solutions are complex and tedious. However, by means
of the graphical inter-relationships between the equilibrium
values of Ca, Alkalinity, Acidity and pH (as depicted in the
Modified Caldwell-Lawrence Diagram) a simple and rapid means of
obtaining solutions is available.

Application requires that the values be known for the equilibrium constants K_1, K_2, K_w and K_s at the different temperature
conditions. Empirical expressions relating the values of K_1, K_2,
K_w and K_s with temperature are given by Eqs. (25 to 27), Chapter
3 and Eq. (4), Chapter 4 respectively.

These relationships were only verified up to 40°C for K_1 and
K_2, and 80°C for K_s. However, values of these constants are required up to about 90°C for solution of water treatment problems
in hot water systems. Langelier, 1946, Powell *et al*, 1945 and
Hamer *et al*, 1961, accept that the values of the equilibrium constants can be extrapolated from the empirical expressions. Langelier states that using extrapolated values for the constants at
100°C the predicted pH values at 100°C are within $\pm$ 0,05 pH units
of the true values.

For our purposes the K values at elevated temperatures given
by the empirical equations will be accepted as satisfactory.

Considering equilibrium between carbonic species in the
aqueous phase, the Caldwell-Lawrence diagram requires knowledge
of two of the parameters Alkalinity, Acidity or pH in order to

define the value of the third. If one has measurements for two
of these parameters at the high temperature, estimation of the
third value from the diagram is straightforward. However, normally the values of Alkalinity and pH only are available at ambient temperature. It is therefore necessary to enquire whether
it is possible to obtain values for two of these parameters at
the elevated temperature from the measured values at ambient
temperature.

Effect of Temperature on C_T, Alkalinity and Acidity

*C_T, Alkalinity and Acidity are all independent of temperature
changes provided neither $CaCO_3$ is precipitated nor CO_2 exchanged
with the air.*

That C_T is independent of temperature (provided neither $CaCO_3$
is precipitated nor CO_2 exchanged with the air) is self-evident.

The independence of Alkalinity and Acidity of temperature,
however, is not obvious. This independence can be demonstrated
as follows:

A solution is prepared by adding X ppm $NaHCO_3$, Y ppm Na_2CO_3
and Z ppm NaOH (all as ppm $CaCO_3$) to CO_2 free distilled water
at say 25°C. Alkalinity of this solution is given by the basic
theory in page 147, i.e.

$$\text{Alkalinity} = (HCO_3^-)_{added} + (CO_3^=)_{added} + (OH^-)_{added} - (H^+)_{added}$$

$$= X + Y + Z - 0 = X + Y + Z$$

The pH which is established in this solution depends on the
concentrations X, Y and Z and on the values of K_1, K_2 and K_w at
25°C. To measure the Alkalinity one would titrate the solution
from its established pH to the pH of its carbonic acid equivalence point. The pH of this equivalence point depends on the
total carbonic species concentration of the solution and also on
the values of K_1, K_2 and K_w (at 25°C in this case).

The effect of heating this water to 90°C is illustrated as follows:

Suppose one prepared the above solution in CO_2 free water at 90°C. From the basic theory given in page 147, Alkalinity in the water at 90°C is

$$\text{Alkalinity at } 90°C = (HCO_3^-)_{added} + (CO_3^=)_{added} + (OH^-) - (H^+)_{added}$$

$$= X + Y + Z + 0 = X + Y + Z$$

i.e.

Alkalinity at 90°C = Alkalinity at 25°C.

However, the pH established in the solution at 90°C will differ from that at 25°C because of the temperature dependence of K_1, K_2 and K_w.

Furthermore, because the carbonic acid equivalence point pH varies slightly with temperature (see Figure 3, Chapter 3), in measuring Alkalinity at 90°C one should theoretically titrate the solution to a slightly different pH value. (For practical purposes this difference is negligible). Thus provided one titrates to the correct carbonic acid equivalence point, temperature has no effect on the Alkalinity value.

The independence of Acidity of temperature is evident from the relationship between Alkalinity, Acidity and C_T, Eq. (54) Chapter 3, i.e.

$$C_T = (\text{Alkalinity} + \text{Acidity})/2$$

i.e.

$$\text{Acidity} = 2C_T - \text{Acidity}$$

Both C_T and Alkalinity are independent of temperature, and from Eq. (54) Chapter 3, it follows that Acidity must also be independent.

From the discussion above it is a simple matter to determine the Alkalinity, Acidity and pH at an elevated temperature knowing the Alkalinity and pH at laboratory temperature.

Knowing pH and Alkalinity from measurements in the laboratory (at say 25°C) the Acidity is immediately available from the Modified Caldwell-Lawrence diagram for 25°C. As neither Alkalinity nor Acidity change with temperature increase, one has the values of these parameters at say 90°C. Knowing the Alkalinity and Acidity at 90°C, the value of pH is immediately available from the Modified Caldwell-Lawrence diagram at 90°C. Water treatment problems at ambient temperatures for utilization at high temperature fall into two categories (i) single aqueous phase equilibria problems and (ii) two phase equilibria problems involving equilibrium between $CaCO_3$ in the solid phase and carbonic species in an aqueous phase.

(i) Single Aqueous Phase Equilibria Problems

This type of problem assumes that pH is governed only by equilibria between dissolved carbonic species in the aqueous phase, i.e. no $CaCO_3$ precipitation occurs nor does CO_2 transfer between air and water take place. Graphical methods for estimating change in pH with temperature are best illustrated by example.

Example

Analysis of a water gives Alkalinity 70 ppm as $CaCO_3$, pH 8,5, temperature 25°C, zero calcium and ionic strength of 0,01. Estimate the pH of this water at 90°C.

This problem is solved using either the Alkalinity-Acidity-pH equilibrium diagram, or the Modified Caldwell-Lawrence diagram.

Referring to the Alkalinity-Acidity-pH diagram for 25°C, Figure 1:

The intersection point of lines representing Alkalinity 70 ppm and pH 8,5, Point 1, defines the Acidity as the value read off the Acidity ordinate, i.e. Acidity at 25°C = 67 ppm as $CaCO_3$.

pH in the water at 90°C is found as follows:

Acidity at 90°C = Acidity at 25°C = 67 ppm as $CaCO_3$

Alkalinity at 90°C = Alkalinity at 25°C = 70 ppm as $CaCO_3$

Referring to the Alkalinity-Acidity-pH equilibrium diagram at 90°C, Figure 2:

Intersection of lines representing Alkalinity 70 ppm and Acidity 67 ppm occurs at Point 1'.

The value of the pH line passing through Point 1' defines the pH in the water at 90°C, i.e. pH 7,95.

The problem is just as simply solved using the Modified Caldwell-Lawrence diagram. Referring to the Caldwell-Lawrence diagram at 25°C, Figure 3:

Intersection of lines representing pH 8,5 and Alkalinity 70 ppm (as $CaCO_3$) occurs at Point A. Acidity ordinate value of Point A defines Acidity in the water at 25°C, i.e. Acidity 67 ppm as $CaCO_3$.

pH at 90°C is found as follows:

Acidity at 90°C = Acidity at 25°C = 67 ppm as $CaCO_3$

Alkalinity at 90°C = Alkalinity at 25°C = 70 ppm as $CaCO_3$

Referring to the Modified Caldwell-Lawrence diagram for 90°C, Figure 4:

Intersection of lines representing Alkalinity 70 ppm and Acidity 67 ppm (both expressed as $CaCO_3$) occurs at Point A' in Figure 4.

The value of the pH line passing through Point A' defines the value for pH of the water at 90°C, i.e. pH 7,95.

(ii) Two Phase Equilibrium Problems

This type of problem considers the effect of temperature change on pH in a water where equilibrium between carbonic species in the aqueous phase and $CaCO_3$ in the solid phase governs the characteristics of the water. Altering the temperature of a water not only changes the pH but also affects the solubility of $CaCO_3$ (see page 125).

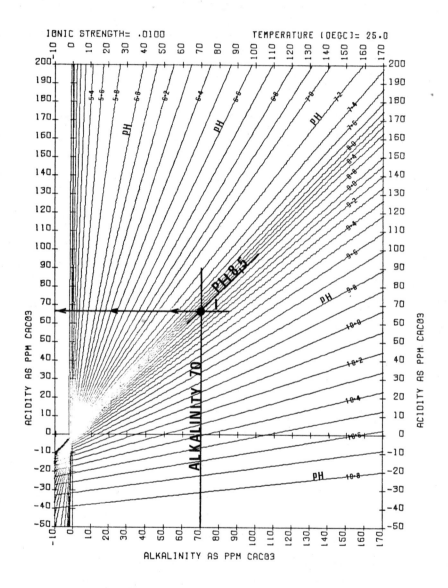

Figure 1.   Initial condition of a water at 25°C is Alkalinity
70 ppm and pH 8,5, i.e. Acidity = 67 ppm

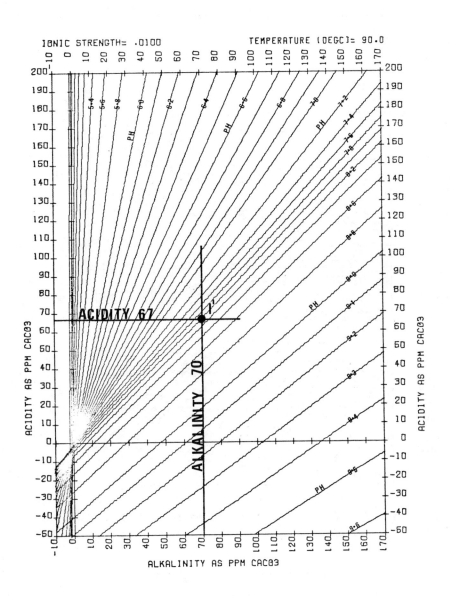

Figure 2. Alkalinity 70 ppm and Acidity 67 ppm gives pH 7,95
at 90°C

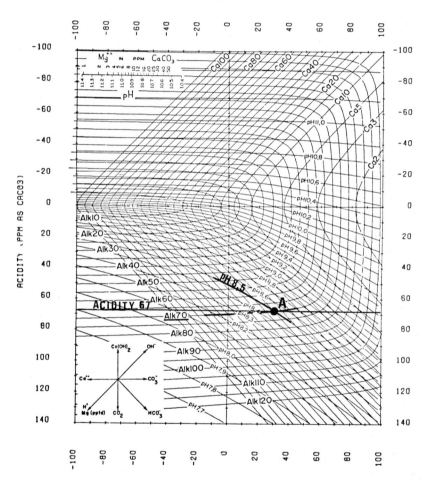

Figure 3. Initial condition of a water at 25°C is Alkalinity 70 ppm and pH 8,5

IONIC STRENGTH= .0100   TEMPERATURE (DEGC)= 90.0

APPROXIMATE TDS(PPM)= 400

Figure 4.   Alkalinity 70 ppm and Acidity 67 ppm gives pH 7,95
at 90°C

Graphical solution procedures for this type of problem are
carried out in two steps: (a) pH in the water at the new temperature is estimated considering the characteristics of the water to
be governed only by equilibrium between carbonic species in the
aqueous phase; and (b) once the pH at the new temperature has
been established, the saturation state with respect to $CaCO_3$ is
estimated in the Modified Caldwell-Lawrence diagram for the relevant temperature using the simple technique described in page
208. Thus, in effect one considers $CaCO_3$ precipitation to be
the result of equilibrium reactions in the aqueous phase which
have already reached completion.

Example

Analysis of a water at 25°C gives Alkalinity 65 ppm, Ca^{++} 65 ppm
(both as $CaCO_3$), pH 8,1 and ionic strength 0,01. Estimate the
following characteristics of this water: (1) the saturation state
with respect to $CaCO_3$ at 25°C; and (2) the saturation state of
the water after heating to 90°C.

1. The initial state of the water at 25°C is shown plotted in
    Figure 5.
    Intersection of the lines representing Alkalinity 65 ppm
    (as $CaCO_3$) and pH 8,1 occurs at Point 1 in Figure 5.
    Acidity ordinate value of Point 1 defines the initial
    Acidity of the water, i.e. Acidity 67 ppm as $CaCO_3$.

    The $CaCO_3$ saturation state of the water is estimated as
    follows:

    The saturated $CaCO_3$ state of the water is given by the
    values of Alkalinity, Ca and pH through Point 2, Figure
    5, i.e. Alkalinity 68 ppm, Ca 68 ppm (both as $CaCO_3$)
    and pH 8,28.

    Comparing the required saturated Alkalinity and Ca values
    of the water, 68 ppm and 68 ppm respectively, with the

actual values, 65 ppm and 65 ppm respectively, one deduces that the water is undersaturated by 3 ppm of $CaCO_3$.

The pH in the water after heating to 90°C is estimated as follows:

Acidity at 90°C = Acidity at 25°C = 67 ppm as $CaCO_3$

Alkalinity at 90°C = Alkalinity at 25°C = 65 ppm as $CaCO_3$

Intersection of the lines representing Acidity 67 ppm and Alkalinity 65 ppm (both as $CaCO_3$) in the Modified Caldwell-Lawrence diagram, plotted for 90°C, occurs at Point 1', Figure 6. The value of the pH line through Point 1' defines the pH in the water after heating to 90°C (provided no $CaCO_3$ is precipitated), i.e. pH 7,80.

2. The saturation state of the water at 90°C is estimated as follows:

The final saturated state which this water ultimately achieves is given by Point 2' in Figure 6. The values of the Alkalinity, Ca and pH lines through Point 2' define the final saturated state of this water, i.e. Alkalinity 58 ppm, Ca 58 ppm (both as $CaCO_3$) and pH 7,43.

Comparing the actual Ca concentration (i.e. 65 ppm) with the final saturated Ca concentration (58 ppm) one deduces that the water is oversaturated by (65 - 58) = 7 ppm of $CaCO_3$ directly after heating.

Thus, there is a potential for 7 ppm of $CaCO_3$ to precipitate from the water after heating to 90°C.

The mass of $CaCO_3$ which has precipitated from the water at any time after heating is found by measuring the pH in the water at 90°C. Plotting this pH value in the Caldwell-Lawrence diagram for 90°C, and noting that

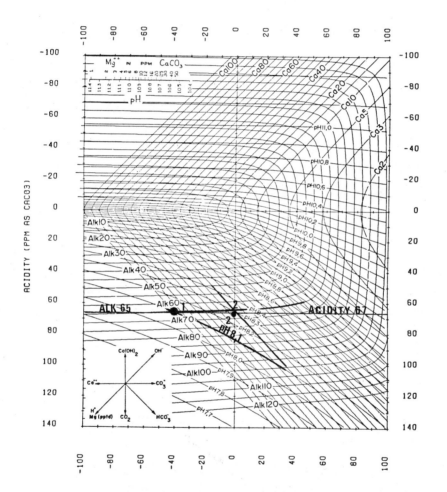

IONIC STRENGTH= .0100   TEMPERATURE (DEGC)= 25.0

APPROXIMATE TDS(PPM)= 400

(ALKALINITY-CALCIUM) (PPM AS CACO3)

Figure 5. Condition of a water at 25°C with Alkalinity 65 ppm, pH 8,1 and Ca 65 ppm – water undersaturated with CaCO$_3$

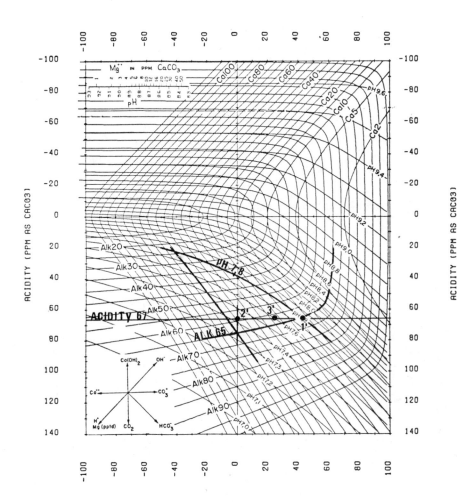

Figure 6. Alkalinity 65 ppm, Acidity 67 ppm and Ca 65 ppm gives
pH 7,80 and CaCO₃ supersaturation at 90°C

Acidity is independent of $CaCO_3$ precipitation, one has an immediate measure of the mass of $CaCO_3$ precipitated.

In the above example suppose the pH in the water is measured as pH 7,60 at 90°C.

Alkalinity in the water is given by the value of the Alkalinity line passing through the intersection point of lines representing pH 7,60 and Acidity 67 ppm, i.e. Alkalinity 62 ppm line through Point 3 in Figure 6.

The concentration of $CaCO_3$ which has precipitated from the water is given by the change in Alkalinity from its over saturated state to the Alkalinity value at pH 7,60 i.e.

$CaCO_3$ precipitated = (65 - 62)

= 3 ppm.

References

Langelier,W.F.   1946   "Effect of Temperature on the pH of Natural Waters", J.A.W.W.A., 38, 178.

Powell, S.T., Bacon, H.E. and Lill, J.R.   1945   "Corrosion Prevention by Controlled Calcium Carbonate Scale, Industrial and Engineering Chemistry, 37, 842.

Hamer, P., Jackson, J. and Thurston, E.F.   1961   *Industrial Water Treatment Practice*, Butterworth, London.

Appendix A

DETERMINATION OF BUFFER CAPACITY (β)

In this appendix tables are presented relating buffer capacity to pH and Alkalinity for a water with some temperature and ionic strength. Calculation of β from these tables is a quick, simple procedure. Calculation of buffer capacity is not practical using those equations in Chapters 2 and 3 linking β to C_T and $[H^+]$. Firstly, C_T is not a readily measurable parameter in water analyses and secondly, solution to these equations requires complex computations.

An equation relating β to pH and Alkalinity is developed as follows:

Equilibrium and mass balance equations are

$$H^+ \cdot HCO_3^-/H_2CO_3^* = K'_{c1} \tag{A1}$$

$$H^+ \cdot CO_3^=/HCO_3^- = K'_{c2} \tag{A2}$$

$$H^+ \cdot OH^- = K'_{cW} \tag{A3}$$

$$Alk = CO_3^= + HCO_3^- + OH - H \tag{A4}$$

$$C_T = H_2CO_3^*/2 + HCO_3^- + CO_3^=/2 \tag{A5}$$

$$= 2{,}303(H^+ + OH^-) + 2{,}303 C_T (K'_{c1} \cdot H^+/(K'_{c1} + H^+)^2$$

$$+ K'_{c2} \cdot H^+/(K'_{c2} + H^+)^2) \tag{A6}$$

Solve for $CO_3^=$ in Eq. (A2), substitute into Eq. (A4) and solve for HCO_3^-:

$$HCO_3^- = (Alk - OH^- + H^+)/(1 + K'_{c2}/H^+) \tag{A7}$$

$$= X$$

Solving for $CO_3^=$ and $H_2CO_3^*$ from Eqs. (A1 and A2) in terms of X (i.e. HCO_3^- and H^+)

$$CO_3^= = K'_{c2}X/H^+ \tag{A8}$$

$$H_2CO_3^* = H^+ . X/K'_{c1} \tag{A9}$$

Substituting Eqs. (A7 to A9) into Eq. (A5)

$$C_T = X(H^+/2K'_{c1} + 1 + K'_{c2}/2H^+) \tag{A10}$$

and substituting Eq. (A10) into the equation for buffer capacity, Eq. (A6), gives an equation linking β to pH and Alkalinity for a water with some temperature and ionic strength.

$$\beta = 2,303(H^+ + OH^-) + X(H^+/2K'_{c1} + 1 + K'_{c2}/2H^+)(2,303\ K'_{c1}\ H^+/ $$
$$(K'_{c1} + H^+)^2 + 2,303\ K'_{c2}\ H^+/(K'_{c2} + H)^2)$$

Substituting for X from Eq. (A7) into the above equation and simplifying

$$\beta = F + Y(Alk - Z) \tag{A11}$$

where

$$F = 2,303(H^+ + OH^-)$$
$$Y = (H^+/2K'_{c1} + 1 + K'_{c2}/2H^+)(2,303.H^+)(K'_{c1}/(K'_{c1} + H^+)^2$$
$$+ K'_{c2}/(K'_{c2} + H^+)^2)/(1 + K'_{c2}/H^+)$$
$$Z = OH^- - H^+$$

In the following tables are listed the factors F, Y and Z (in Eq. (A11)) versus pH for water with some measured temperature and ionic strength. These tables can now be used for rapidly assessing the buffer capacity of a water.

To illustrate the use of the buffer capacity tables a comparison of buffer capacities is made of the final water qualities obtained from the two treatment methods proposed in Chapter 6 (see pages 265 to 270):

(i)   Alkalinity = 30 ppm as $CaCO_3$, pH = 9,0 and S.I. = +0,2
      (attained using 28 ppm $Ca(OH)_2$ and recarbonating with
      CO_2) and

(ii) Alkalinity = 13 ppm as $CaCO_3$, pH = 10,1 and S.I. = +0,2
      (attained using 12 ppm $Ca(OH)_2$)

      (Assume water temperature of 20°C and μ = 0,001).

(i)   From the table for temperature 20°C and β = 0,001 at
      pH 9,0: F = 0,806   Y = 0,099   Z = 0,35

      β for Alk 30 (ppm as $CaCO_3$) and pH 9,0 is now calcu-
      lated from Eq. (A11),

      β = F + Y (Alk - Z)

        = 0,806 + 0,099 (30 - 0,35)

        = 3,74 ppm as $CaCO_3$/unit pH.

(ii) From the relevant table

      at pH 10,1: F = 10,1   Y = 0,392   Z = 4,4   and

      β = 10,1 + 0,392 (13 - 4,4)

        = 13,47 ppm as $CaCO_3$/unit pH.

Comparing these two waters: adjusting the water to pH 10,1
gives a buffer capacity 3,6 times that for the pH 9,0 and Alk 30
proposal; lime dosage to pH 10,1 is less than a half that required to pH 9,0 and no recarbonation is required.

BUFFER CAPACITY TABLES

TEMP(DEGC)=10.0   IONIC STRENGTH= .0010
BUFFER CAPACITY=F+Y(ALK.-Z)
BUFFER CAPACITY IN PPM AS CACO3/PH
ALKALINITY IN PPM AS CACO3

BUFFER CAPACITY FACTORS:F,Y,Z

PH	F	Y	Z
.500+01	.119+01	.223+01	-.518+00
.520+01	.753+00	.218+01	-.327+00
.540+01	.475+00	.212+01	-.206+00
.560+01	.300+00	.202+01	-.130+00
.580+01	.189+00	.189+01	-.820-01
.600+01	.120+00	.171+01	-.516-01
.620+01	.758-01	.148+01	-.324-01
.640+01	.483-01	.123+01	-.202-01
.660+01	.313-01	.965+00	-.124-01
.680+01	.211-01	.721+00	-.726-02
.700+01	.154-01	.514+00	-.367-02
.720+01	.130-01	.354+00	-.880-03
.740+01	.135-01	.238+00	.172-02
.760+01	.168-01	.158+00	.470-02
.780+01	.238-01	.105+00	.868-02
.800+01	.359-01	.724-01	.145-01
.820+01	.557-01	.538-01	.235-01
.840+01	.876-01	.462-01	.376-01
.860+01	.138+00	.480-01	.598-01
.880+01	.219+00	.592-01	.949-01
.900+01	.347+00	.811-01	.151+00
.920+01	.550+00	.116+00	.239+00
.940+01	.871+00	.164+00	.378+00
.960+01	.138+01	.226+00	.600+00
.980+01	.219+01	.294+00	.950+00
.100+02	.347+01	.355+00	.151+01
.102+02	.550+01	.391+00	.239+01
.104+02	.871+01	.390+00	.378+01
.106+02	.138+02	.351+00	.600+01
.108+02	.219+02	.289+00	.950+01
.110+02	.347+02	.220+00	.151+02

BUFFER CAPACITY TABLES

TEMP(DEGC)=10.0 IONIC STRENGTH= .0025
BUFFER CAPACITY=F+Y(ALK.-Z)
BUFFER CAPACITY IN PPM AS CACO3/PH
ALKALINITY IN PPM AS CACO3

BUFFER CAPACITY FACTORS:F,Y,Z

PH	F	Y	Z
.500+01	.122+01	.222+01	-.528+00
.520+01	.767+00	.218+01	-.333+00
.540+01	.484+00	.211+01	-.210+00
.560+01	.306+00	.202+01	-.133+00
.580+01	.193+00	.188+01	-.836-01
.600+01	.122+00	.170+01	-.526-01
.620+01	.773-01	.147+01	-.331-01
.640+01	.493-01	.122+01	-.206-01
.660+01	.319-01	.954+00	-.126-01
.680+01	.215-01	.711+00	-.740-02
.700+01	.157-01	.507+00	-.374-02
.720+01	.133-01	.349+00	-.897-03
.740+01	.137-01	.234+00	.175-02
.760+01	.171-01	.156+00	.479-02
.780+01	.242-01	.104+00	.885-02
.800+01	.366-01	.717-01	.148-01
.820+01	.568-01	.538-01	.240-01
.840+01	.893-01	.469-01	.384-01
.860+01	.141+00	.495-01	.610-01
.880+01	.223+00	.617-01	.968-01
.900+01	.354+00	.849-01	.153+00
.920+01	.560+00	.121+00	.243+00
.940+01	.888+00	.171+00	.386+00
.960+01	.141+01	.234+00	.611+00
.980+01	.223+01	.302+00	.969+00
.100+02	.354+01	.361+00	.154+01
.102+02	.560+01	.393+00	.243+01
.104+02	.888+01	.387+00	.386+01
.106+02	.141+02	.344+00	.611+01
.108+02	.223+02	.280+00	.969+01
.110+02	.354+02	.212+00	.154+02

BUFFER CAPACITY TABLES

TEMP(DEGC)=10.0   IONIC STRENGTH= .0050
BUFFER CAPACITY=F+Y(ALK.-Z)
BUFFER CAPACITY IN PPM AS CACO3/PH
ALKALINITY IN PPM AS CACO3

BUFFER CAPACITY FACTORS:F,Y,Z

PH	F	Y	Z
.500+01	.124+01	.222+01	-.539+00
.520+01	.783+00	.218+01	-.340+00
.540+01	.494+00	.211+01	-.214+00
.560+01	.312+00	.201+01	-.135+00
.580+01	.197+00	.187+01	-.853-01
.600+01	.124+00	.169+01	-.537-01
.620+01	.789-01	.146+01	-.338-01
.640+01	.503-01	.121+01	-.211-01
.660+01	.326-01	.943+00	-.129-01
.680+01	.219-01	.701+00	-.755-02
.700+01	.160-01	.499+00	-.382-02
.720+01	.136-01	.343+00	-.916-03
.740+01	.140-01	.230+00	.179-02
.760+01	.175-01	.153+00	.489-02
.780+01	.247-01	.102+00	.903-02
.800+01	.373-01	.710-01	.151-01
.820+01	.580-01	.539-01	.245-01
.840+01	.912-01	.477-01	.392-01
.860+01	.144+00	.512-01	.623-01
.880+01	.228+00	.646-01	.988-01
.900+01	.361+00	.892-01	.157+00
.920+01	.572+00	.127+00	.248+00
.940+01	.907+00	.179+00	.394+00
.960+01	.144+01	.243+00	.624+00
.980+01	.228+01	.311+00	.989+00
.100+02	.361+01	.367+00	.157+01
.102+02	.572+01	.395+00	.248+01
.104+02	.907+01	.383+00	.394+01
.106+02	.144+02	.337+00	.624+01
.108+02	.228+02	.271+00	.989+01
.110+02	.361+02	.203+00	.157+02

BUFFER CAPACITY TABLES

TEMP(DEGC)=10.0   IONIC STRENGTH= .0075
BUFFER CAPACITY=F+Y(ALK.-Z)
BUFFER CAPACITY IN PPM AS CACO3/PH
ALKALINITY IN PPM AS CACO3

BUFFER CAPACITY FACTORS:F,Y,Z

PH	F	Y	Z
.500+01	.126+01	.222+01	-.547+00
.520+01	.795+00	.218+01	-.345+00
.540+01	.502+00	.211+01	-.218+00
.560+01	.317+00	.201+01	-.137+00
.580+01	.200+00	.187+01	-.866-01
.600+01	.126+00	.168+01	-.546-01
.620+01	.801-01	.145+01	-.343-01
.640+01	.511-01	.120+01	-.214-01
.660+01	.331-01	.935+00	-.131-01
.680+01	.223-01	.694+00	-.767-02
.700+01	.163-01	.493+00	-.388-02
.720+01	.138-01	.339+00	-.930-03
.740+01	.142-01	.227+00	.182-02
.760+01	.178-01	.151+00	.496-02
.780+01	.251-01	.101+00	.917-02
.800+01	.379-01	.705-01	.154-01
.820+01	.589-01	.540-01	.249-01
.840+01	.925-01	.484-01	.398-01
.860+01	.146+00	.525-01	.632-01
.880+01	.231+00	.668-01	.100+00
.900+01	.367+00	.925-01	.159+00
.920+01	.581+00	.132+00	.252+00
.940+01	.921+00	.185+00	.400+00
.960+01	.146+01	.250+00	.633+00
.980+01	.231+01	.317+00	.100+01
.100+02	.366+01	.372+00	.159+01
.102+02	.581+01	.395+00	.252+01
.104+02	.920+01	.380+00	.400+01
.106+02	.146+02	.331+00	.633+01
.108+02	.231+02	.264+00	.100+02
.110+02	.366+02	.197+00	.159+02

BUFFER CAPACITY TABLES

TEMP(DEGC)=10.0   IONIC STRENGTH= .0100
BUFFER CAPACITY=F+Y(ALK.-Z)
BUFFER CAPACITY IN PPM AS CACO3/PH
ALKALINITY IN PPM AS CACO3

BUFFER CAPACITY FACTORS:F,Y,Z

PH	F	Y	Z
.500+01	.128+01	.222+01	-.554+00
.520+01	.805+00	.217+01	-.349+00
.540+01	.508+00	.211+01	-.220+00
.560+01	.321+00	.201+01	-.139+00
.580+01	.202+00	.186+01	-.877-01
.600+01	.128+00	.168+01	-.552-01
.620+01	.811-01	.145+01	-.347-01
.640+01	.517-01	.119+01	-.216-01
.660+01	.335-01	.928+00	-.133-01
.680+01	.226-01	.688+00	-.776-02
.700+01	.165-01	.488+00	-.393-02
.720+01	.139-01	.335+00	-.942-03
.740+01	.144-01	.225+00	.184-02
.760+01	.180-01	.149+00	.502-02
.780+01	.254-01	.100+00	.929-02
.800+01	.384-01	.702-01	.156-01
.820+01	.596-01	.541-01	.252-01
.840+01	.937-01	.489-01	.402-01
.860+01	.148+00	.537-01	.640-01
.880+01	.234+00	.686-01	.102+00
.900+01	.371+00	.953-01	.161+00
.920+01	.588+00	.136+00	.255+00
.940+01	.932+00	.190+00	.405+00
.960+01	.148+01	.256+00	.641+00
.980+01	.234+01	.323+00	.102+01
.100+02	.371+01	.375+00	.161+01
.102+02	.588+01	.395+00	.255+01
.104+02	.932+01	.377+00	.405+01
.136+02	.148+02	.326+00	.641+01
.108+02	.234+02	.259+00	.102+02
.110+02	.371+02	.192+00	.161+02

BUFFER CAPACITY TABLES

TEMP(DEGC)=10.0   IONIC STRENGTH= .0150
BUFFER CAPACITY=F+Y(ALK.-Z)
BUFFER CAPACITY IN PPM AS CACO3/PH
ALKALINITY IN PPM AS CACO3

BUFFER CAPACITY FACTORS:F,Y,Z

PH	F	Y	Z
.500+01	.130+01	.222+01	-.565+00
.520+01	.821+00	.217+01	-.356+00
.540+01	.518+00	.210+01	-.225+00
.560+01	.327+00	.200+01	-.142+00
.580+01	.206+00	.186+01	-.894-01
.600+01	.130+00	.167+01	-.563-01
.620+01	.827-01	.144+01	-.354-01
.640+01	.527-01	.118+01	-.221-01
.660+01	.342-01	.917+00	-.135-01
.680+01	.230-01	.678+00	-.792-02
.700+01	.168-01	.481+00	-.401-02
.720+01	.142-01	.330+00	-.961-03
.740+01	.147-01	.221+00	.188-02
.760+01	.183-01	.147+00	.512-02
.780+01	.259-01	.989-01	.947-02
.800+01	.391-01	.696-01	.159-01
.820+01	.608-01	.543-01	.257-01
.840+01	.956-01	.499-01	.410-01
.860+01	.151+00	.556-01	.653-01
.880+01	.239+00	.717-01	.104+00
.900+01	.379+00	.999-01	.164+00
.920+01	.600+00	.142+00	.260+00
.940+01	.951+00	.198+00	.413+00
.960+01	.151+01	.264+00	.654+00
.980+01	.239+01	.331+00	.104+01
.100+02	.378+01	.380+00	.164+01
.102+02	.600+01	.395+00	.260+01
.104+02	.951+01	.372+00	.413+01
.106+02	.151+02	.318+00	.654+01
.108+02	.239+02	.250+00	.104+02
.110+02	.378+02	.184+00	.164+02

BUFFER CAPACITY TABLES

TEMP(DEGC)=15.0   IONIC STRENGTH= .0010
BUFFER CAPACITY=F+Y(ALK.-Z)
BUFFER CAPACITY IN PPM AS CACO3/PH
ALKALINITY IN PPM AS CACO3

BUFFER CAPACITY FACTORS:F,Y,Z

PH	F	Y	Z
.500+01	.119+01	.222+01	-.518+00
.520+01	.753+00	.217+01	-.327+00
.540+01	.475+00	.210+01	-.206+00
.560+01	.300+00	.200+01	-.130+00
.580+01	.189+00	.185+01	-.819-01
.600+01	.120+00	.166+01	-.516-01
.620+01	.761-01	.143+01	-.323-01
.640+01	.488-01	.117+01	-.200-01
.660+01	.321-01	.910+00	-.121-01
.680+01	.223-01	.673+00	-.674-02
.700+01	.173-01	.476+00	-.286-02
.720+01	.160-01	.326+00	.409-03
.740+01	.182-01	.218+00	.377-02
.760+01	.243-01	.145+00	.793-02
.780+01	.356-01	.971-01	.138-01
.800+01	.546-01	.677-01	.227-01
.820+01	.854-01	.519-01	.364-01
.840+01	.135+00	.467-01	.581-01
.860+01	.213+00	.509-01	.922-01
.880+01	.337+00	.648-01	.146+00
.900+01	.534+00	.901-01	.232+00
.920+01	.847+00	.128+00	.368+00
.940+01	.134+01	.181+00	.583+00
.960+01	.213+01	.246+00	.923+00
.980+01	.337+01	.313+00	.146+01
.100+02	.534+01	.369+00	.232+01
.102+02	.847+01	.395+00	.368+01
.104+02	.134+02	.382+00	.583+01
.106+02	.213+02	.334+00	.923+01
.108+02	.337+02	.269+00	.146+02
.110+02	.534+02	.201+00	.232+02

BUFFER CAPACITY TABLES

TEMP(DEGC)=15.0   IONIC STRENGTH= .0025
BUFFER CAPACITY=F+Y(ALK.-Z)
BUFFER CAPACITY IN PPM AS CACO3/PH
ALKALINITY IN PPM AS CACO3

BUFFER CAPACITY FACTORS:F,Y,Z

PH	F	Y	Z
.500+01	.122+01	.222+01	-.528+00
.520+01	.767+00	.217+01	-.333+00
.540+01	.484+00	.210+01	-.210+00
.560+01	.306+00	.199+01	-.133+00
.580+01	.193+00	.185+01	-.835-01
.600+01	.122+00	.165+01	-.526-01
.620+01	.776-01	.142+01	-.329-01
.640+01	.498-01	.116+01	-.204-01
.660+01	.327-01	.900+00	-.123-01
.680+01	.227-01	.664+00	-.687-02
.700+01	.176-01	.469+00	-.291-02
.720+01	.163-01	.321+00	.417-03
.740+01	.185-01	.215+00	.384-02
.760+01	.247-01	.143+00	.809-02
.780+01	.363-01	.958-01	.141-01
.800+01	.557-01	.672-01	.231-01
.820+01	.871-01	.521-01	.371-01
.840+01	.137+00	.476-01	.592-01
.860+01	.217+00	.526-01	.940-01
.880+01	.344+00	.677-01	.149+00
.900+01	.545+00	.943-01	.236+00
.920+01	.863+00	.134+00	.375+00
.940+01	.137+01	.189+00	.594+00
.960+01	.217+01	.254+00	.941+00
.980+01	.344+01	.321+00	.149+01
.100+02	.545+01	.374+00	.236+01
.102+02	.863+01	.395+00	.375+01
.104+02	.137+02	.378+00	.594+01
.106+02	.217+02	.327+00	.941+01
.108+02	.344+02	.260+00	.149+02
.110+02	.545+02	.193+00	.236+02

BUFFER CAPACITY TABLES

TEMP(DEGC)=15.0   IONIC STRENGTH= .0050
BUFFER CAPACITY=F+Y(ALK.-Z)
BUFFER CAPACITY IN PPM AS CACO3/PH
ALKALINITY IN PPM AS CACO3

BUFFER CAPACITY FACTORS:F,Y,Z

PH	F	Y	Z
.500+01	.124+01	.221+01	-.539+00
.520+01	.783+00	.217+01	-.340+00
.540+01	.494+00	.209+01	-.214+00
.560+01	.312+00	.199+01	-.135+00
.580+01	.197+00	.184+01	-.853-01
.600+01	.125+00	.164+01	-.536-01
.620+01	.792-01	.141+01	-.336-01
.640+01	.508-01	.115+01	-.208-01
.660+01	.334-01	.888+00	-.126-01
.680+01	.232-01	.654+00	-.702-02
.700+01	.180-01	.461+00	-.297-02
.720+01	.166-01	.315+00	.426-03
.740+01	.189-01	.211+00	.392-02
.760+01	.252-01	.140+00	.826-02
.780+01	.370-01	.944-01	.144-01
.800+01	.568-01	.667-01	.236-01
.820+01	.889-01	.523-01	.379-01
.840+01	.140+00	.486-01	.604-01
.860+01	.222+00	.546-01	.960-01
.880+01	.351+00	.709-01	.152+00
.900+01	.556+00	.991-01	.241+00
.920+01	.881+00	.141+00	.383+00
.940+01	.140+01	.197+00	.606+00
.960+01	.221+01	.264+00	.961+00
.980+01	.351+01	.330+00	.152+01
.100+02	.556+01	.379+00	.241+01
.102+02	.881+01	.395+00	.383+01
.104+02	.140+02	.372+00	.606+01
.106+02	.221+02	.318+00	.961+01
.108+02	.351+02	.251+00	.152+02
.110+02	.556+02	.184+00	.241+02

BUFFER CAPACITY TABLES

TEMP(DEGC)=15.0   IONIC STRENGTH= .0075
BUFFER CAPACITY=F+Y(ALK.-Z)
BUFFER CAPACITY IN PPM AS CACO3/PH
ALKALINITY IN PPM AS CACO3

BUFFER CAPACITY FACTORS:F,Y,Z

PH	F	Y	Z
.500+01	.126+01	.221+01	-.547+00
.520+01	.795+00	.216+01	-.345+00
.540+01	.502+00	.209+01	-.218+00
.560+01	.317+00	.198+01	-.137+00
.580+01	.200+00	.183+01	-.866-01
.600+01	.127+00	.164+01	-.545-01
.620+01	.804-01	.140+01	-.341-01
.640+01	.516-01	.114+01	-.212-01
.660+01	.339-01	.880+00	-.128-01
.680+01	.235-01	.647+00	-.712-02
.700+01	.182-01	.456+00	-.302-02
.720+01	.169-01	.311+00	.432-03
.740+01	.192-01	.208+00	.398-02
.760+01	.256-01	.138+00	.838-02
.780+01	.376-01	.934-01	.146-01
.800+01	.577-01	.663-01	.240-01
.820+01	.902-01	.525-01	.385-01
.840+01	.142+00	.494-01	.613-01
.860+01	.225+00	.562-01	.974-01
.880+01	.356+00	.734-01	.155+00
.900+01	.565+00	.103+00	.245+00
.920+01	.895+00	.146+00	.388+00
.940+01	.142+01	.203+00	.616+00
.960+01	.225+01	.270+00	.976+00
.980+01	.356+01	.336+00	.155+01
.100+02	.564+01	.382+00	.245+01
.102+02	.895+01	.395+00	.388+01
.104+02	.142+02	.368+00	.616+01
.106+02	.225+02	.312+00	.976+01
.108+02	.356+02	.244+00	.155+02
.110+02	.564+02	.178+00	.245+02

BUFFER CAPACITY TABLES

TEMP(DEGC)=15.0   IONIC STRENGTH= .0100
BUFFER CAPACITY=F+Y(ALK.-Z)
BUFFER CAPACITY IN PPM AS CACO3/PH
ALKALINITY IN PPM AS CACO3

BUFFER CAPACITY FACTORS:F,Y,Z

PH	F	Y	Z
.500+01	.128+01	.221+01	-.554+00
.520+01	.805+00	.216+01	-.349+00
.540+01	.508+00	.209+01	-.220+00
.560+01	.321+00	.198+01	-.139+00
.580+01	.203+00	.183+01	-.876-01
.600+01	.128+00	.163+01	-.551-01
.620+01	.814-01	.139+01	-.346-01
.640+01	.522-01	.113+01	-.214-01
.660+01	.343-01	.874+00	-.129-01
.680+01	.238-01	.641+00	-.721-02
.700+01	.185-01	.451+00	-.306-02
.720+01	.171-01	.308+00	.438-03
.740+01	.194-01	.206+00	.403-02
.760+01	.260-01	.137+00	.849-02
.780+01	.381-01	.927-01	.148-01
.800+01	.584-01	.661-01	.243-01
.820+01	.914-01	.527-01	.390-01
.840+01	.144+00	.501-01	.621-01
.860+01	.228+00	.575-01	.986-01
.880+01	.361+00	.755-01	.156+00
.900+01	.572+00	.106+00	.248+00
.920+01	.906+00	.150+00	.393+00
.940+01	.144+01	.209+00	.623+00
.960+01	.227+01	.276+00	.988+00
.980+01	.361+01	.340+00	.157+01
.100+02	.571+01	.385+00	.248+01
.102+02	.906+01	.394+00	.393+01
.104+02	.144+02	.365+00	.623+01
.106+02	.227+02	.307+00	.988+01
.108+02	.361+02	.238+00	.157+02
.110+02	.571+02	.174+00	.248+02

BUFFER CAPACITY TABLES

TEMP(DEGC)=15.0   IONIC STRENGTH= .0150
BUFFER CAPACITY=F+Y(ALK.-Z)
BUFFER CAPACITY IN PPM AS CACO3/PH
ALKALINITY IN PPM AS CACO3

BUFFER CAPACITY FACTORS:F,Y,Z

PH	F	Y	Z
.500+01	.130+01	.221+01	-.565+00
.520+01	.821+00	.216+01	-.356+00
.540+01	.518+00	.208+01	-.225+00
.560+01	.327+00	.197+01	-.142+00
.580+01	.207+00	.182+01	-.894-01
.600+01	.131+00	.162+01	-.562-01
.620+01	.830-01	.138+01	-.352-01
.640+01	.533-01	.112+01	-.219-01
.660+01	.350-01	.863+00	-.132-01
.680+01	.243-01	.632+00	-.736-02
.700+01	.188-01	.444+00	-.312-02
.720+01	.174-01	.303+00	.446-03
.740+01	.198-01	.203+00	.411-02
.760+01	.265-01	.135+00	.866-02
.780+01	.388-01	.915-01	.151-01
.800+01	.596-01	.657-01	.247-01
.820+01	.932-01	.531-01	.398-01
.840+01	.147+00	.513-01	.633-01
.860+01	.232+00	.597-01	.101+00
.880+01	.368+00	.790-01	.160+00
.900+01	.583+00	.111+00	.253+00
.920+01	.924+00	.157+00	.401+00
.940+01	.146+01	.217+00	.636+00
.960+01	.232+01	.285+00	.101+01
.980+01	.368+01	.348+00	.160+01
.100+02	.583+01	.388+00	.253+01
.102+02	.924+01	.392+00	.401+01
.104+02	.146+02	.358+00	.636+01
.106+02	.232+02	.298+00	.101+02
.108+02	.368+02	.229+00	.160+02
.110+02	.583+02	.166+00	.253+02

BUFFER CAPACITY TABLES

TEMP(DEGC)=20.0   IONIC STRENGTH= .0010
BUFFER CAPACITY=F+Y(ALK.-Z)
BUFFER CAPACITY IN PPM AS CACO3/PH
ALKALINITY IN PPM AS CACO3

BUFFER CAPACITY FACTORS:F,Y,Z

PH	F	Y	Z
.500+01	.119+01	.221+01	-.518+00
.520+01	.753+00	.216+01	-.327+00
.540+01	.475+00	.208+01	-.206+00
.560+01	.300+00	.198+01	-.130+00
.580+01	.190+00	.182+01	-.819-01
.600+01	.120+00	.163+01	-.514-01
.620+01	.765-01	.139+01	-.321-01
.640+01	.495-01	.112+01	-.197-01
.660+01	.332-01	.866+00	-.116-01
.680+01	.240-01	.635+00	-.600-02
.700+01	.200-01	.446+00	-.168-02
.720+01	.203-01	.304+00	.228-02
.740+01	.250-01	.203+00	.673-02
.760+01	.351-01	.135+00	.126-01
.780+01	.527-01	.910-01	.213-01
.800+01	.818-01	.645-01	.345-01
.820+01	.128+00	.509-01	.551-01
.840+01	.203+00	.477-01	.877-01
.860+01	.321+00	.541-01	.139+00
.880+01	.509+00	.707-01	.221+00
.900+01	.806+00	.991-01	.350+00
.920+01	.128+01	.141+00	.555+00
.940+01	.202+01	.197+00	.879+00
.960+01	.321+01	.264+00	.139+01
.980+01	.508+01	.330+00	.221+01
.100+02	.806+01	.379+00	.350+01
.102+02	.128+02	.395+00	.555+01
.104+02	.202+02	.372+00	.879+01
.106+02	.321+02	.318+00	.139+02
.108+02	.508+02	.250+00	.221+02
.110+02	.806+02	.184+00	.350+02

BUFFER CAPACITY TABLES

TEMP(DEGC)=20.0   IONIC STRENGTH= .0025
BUFFER CAPACITY=F+Y(ALK.-Z)
BUFFER CAPACITY IN PPM AS CACO3/PH
ALKALINITY IN PPM AS CACO3

BUFFER CAPACITY FACTORS:F,Y,Z

PH	F	Y	Z
.500+01	.122+01	.221+01	-.528+00
.520+01	.767+00	.216+01	-.333+00
.540+01	.484+00	.208+01	-.210+00
.560+01	.306+00	.197+01	-.132+00
.580+01	.193+00	.182+01	-.834-01
.600+01	.122+00	.162+01	-.524-01
.620+01	.780-01	.138+01	-.327-01
.640+01	.505-01	.111+01	-.201-01
.660+01	.338-01	.856+00	-.118-01
.680+01	.245-01	.626+00	-.612-02
.700+01	.204-01	.439+00	-.171-02
.720+01	.207-01	.299+00	.232-02
.740+01	.255-01	.200+00	.686-02
.760+01	.358-01	.133+00	.129-01
.780+01	.538-01	.898-01	.217-01
.800+01	.834-01	.641-01	.351-01
.820+01	.131+00	.512-01	.562-01
.840+01	.207+00	.488-01	.894-01
.860+01	.327+00	.561-01	.142+00
.880+01	.518+00	.739-01	.225+00
.900+01	.822+00	.104+00	.357+00
.920+01	.130+01	.148+00	.565+00
.940+01	.206+01	.205+00	.896+00
.960+01	.327+01	.272+00	.142+01
.980+01	.518+01	.338+00	.225+01
.100+02	.821+01	.383+00	.357+01
.102+02	.130+02	.394+00	.565+01
.104+02	.206+02	.367+00	.896+01
.106+02	.327+02	.310+00	.142+02
.108+02	.518+02	.242+00	.225+02
.110+02	.821+02	.176+00	.357+02

BUFFER CAPACITY TABLES

TEMP(DEGC)=20.0   IONIC STRENGTH= .0050
BUFFER CAPACITY=F+Y(ALK.-Z)
BUFFER CAPACITY IN PPM AS CACO3/PH
ALKALINITY IN PPM AS CACO3

BUFFER CAPACITY FACTORS:F,Y,Z

PH	F	Y	Z
.500+01	.124+01	.221+01	-.539+00
.520+01	.783+00	.215+01	-.340+00
.540+01	.494+00	.208+01	-.214+00
.560+01	.312+00	.196+01	-.135+00
.580+01	.197+00	.181+01	-.852-01
.600+01	.125+00	.161+01	-.535-01
.620+01	.796-01	.136+01	-.334-01
.640+01	.515-01	.110+01	-.205-01
.660+01	.345-01	.844+00	-.121-01
.680+01	.250-01	.617+00	-.624-02
.700+01	.208-01	.432+00	-.175-02
.720+01	.211-01	.294+00	.237-02
.740+01	.260-01	.196+00	.700-02
.760+01	.365-01	.131+00	.131-01
.780+01	.549-01	.886-01	.221-01
.800+01	.851-01	.637-01	.359-01
.820+01	.134+00	.516-01	.574-01
.840+01	.211+00	.501-01	.912-01
.860+01	.334+00	.584-01	.145+00
.880+01	.529+00	.775-01	.230+00
.900+01	.839+00	.109+00	.364+00
.920+01	.133+01	.155+00	.577+00
.940+01	.211+01	.214+00	.915+00
.960+01	.334+01	.282+00	.145+01
.980+01	.529+01	.345+00	.230+01
.100+02	.839+01	.387+00	.364+01
.102+02	.133+02	.393+00	.577+01
.104+02	.211+02	.360+00	.915+01
.106+02	.334+02	.301+00	.145+02
.108+02	.529+02	.232+00	.230+02
.110+02	.839+02	.168+00	.364+02

BUFFER CAPACITY TABLES

TEMP(DEGC)=20.0   IONIC STRENGTH= .0075
BUFFER CAPACITY=F+Y(ALK.-Z)
BUFFER CAPACITY IN PPM AS CACO3/PH
ALKALINITY IN PPM AS CACO3

BUFFER CAPACITY FACTORS:F,Y,Z

PH	F	Y	Z
.500+01	.126+01	.221+01	-.547+00
.520+01	.795+00	.215+01	-.345+00
.540+01	.502+00	.207+01	-.218+00
.560+01	.317+00	.196+01	-.137+00
.580+01	.200+00	.180+01	-.865-01
.600+01	.127+00	.160+01	-.543-01
.620+01	.808-01	.136+01	-.339-01
.640+01	.523-01	.109+01	-.209-01
.660+01	.350-01	.836+00	-.123-01
.680+01	.253-01	.610+00	-.634-02
.700+01	.211-01	.427+00	-.177-02
.720+01	.214-01	.290+00	.241-02
.740+01	.264-01	.194+00	.711-02
.760+01	.371-01	.129+00	.133-01
.780+01	.557-01	.878-01	.225-01
.800+01	.864-01	.635-01	.364-01
.820+01	.136+00	.520-01	.582-01
.840+01	.214+00	.511-01	.926-01
.860+01	.339+00	.602-01	.147+00
.880+01	.537+00	.803-01	.233+00
.900+01	.851+00	.113+00	.370+00
.920+01	.135+01	.160+00	.586+00
.940+01	.214+01	.221+00	.929+00
.960+01	.339+01	.288+00	.147+01
.980+01	.537+01	.351+00	.233+01
.100+02	.851+01	.389+00	.370+01
.102+02	.135+02	.391+00	.586+01
.104+02	.214+02	.356+00	.929+01
.106+02	.339+02	.294+00	.147+02
.108+02	.537+02	.226+00	.233+02
.110+02	.851+02	.163+00	.370+02

BUFFER CAPACITY TABLES

TEMP(DEGC)=20.0   IONIC STRENGTH= .0100
BUFFER CAPACITY=F+Y(ALK.-Z)
BUFFER CAPACITY IN PPM AS CACO3/PH
ALKALINITY IN PPM AS CACO3

BUFFER CAPACITY FACTORS:F,Y,Z

PH	F	Y	Z
.500+01	.128+01	.220+01	-.554+00
.520+01	.805+00	.215+01	-.349+00
.540+01	.508+00	.207+01	-.220+00
.560+01	.321+00	.196+01	-.139+00
.580+01	.203+00	.180+01	-.875-01
.600+01	.128+00	.159+01	-.550-01
.620+01	.819-01	.135+01	-.344-01
.640+01	.529-01	.109+01	-.211-01
.660+01	.355-01	.830+00	-.124-01
.680+01	.257-01	.604+00	-.642-02
.700+01	.214-01	.423+00	-.180-02
.720+01	.217-01	.287+00	.244-02
.740+01	.267-01	.192+00	.720-02
.760+01	.375-01	.128+00	.135-01
.780+01	.564-01	.871-01	.227-01
.800+01	.875-01	.633-01	.369-01
.820+01	.137+00	.523-01	.590-01
.840+01	.217+00	.519-01	.938-01
.860+01	.343+00	.617-01	.149+00
.880+01	.544+00	.826-01	.236+00
.900+01	.862+00	.116+00	.374+00
.920+01	.137+01	.165+00	.593+00
.940+01	.217+01	.226+00	.940+00
.960+01	.343+01	.294+00	.149+01
.980+01	.544+01	.355+00	.236+01
.100+02	.862+01	.391+00	.374+01
.102+02	.137+02	.390+00	.593+01
.104+02	.216+02	.351+00	.940+01
.106+02	.343+02	.289+00	.149+02
.108+02	.544+02	.220+00	.236+02
.110+02	.862+02	.158+00	.374+02

BUFFER CAPACITY TABLES

TEMP(DEGC)=20.0   IONIC STRENGTH= .0150
BUFFER CAPACITY=F+Y(ALK.-Z)
BUFFER CAPACITY IN PPM AS CACO3/PH
ALKALINITY IN PPM AS CACO3

BUFFER CAPACITY FACTORS:F,Y,Z

PH	F	Y	Z
.500+01	.130+01	.220+01	-.565+00
.520+01	.821+00	.215+01	-.356+00
.540+01	.518+00	.207+01	-.225+00
.560+01	.327+00	.195+01	-.142+00
.580+01	.207+00	.179+01	-.893-01
.600+01	.131+00	.158+01	-.561-01
.620+01	.835-01	.134+01	-.350-01
.640+01	.540-01	.107+01	-.215-01
.660+01	.362-01	.819+00	-.127-01
.680+01	.262-01	.596+00	-.655-02
.700+01	.218-01	.416+00	-.183-02
.720+01	.221-01	.283+00	.249-02
.740+01	.273-01	.189+00	.734-02
.760+01	.383-01	.126+00	.138-01
.780+01	.575-01	.861-01	.232-01
.800+01	.892-01	.632-01	.376-01
.820+01	.140+00	.529-01	.601-01
.840+01	.221+00	.534-01	.957-01
.860+01	.350+00	.642-01	.152+00
.880+01	.555+00	.865-01	.241+00
.900+01	.879+00	.122+00	.382+00
.920+01	.139+01	.172+00	.605+00
.940+01	.221+01	.235+00	.959+00
.960+01	.350+01	.303+00	.152+01
.980+01	.555+01	.361+00	.241+01
.100+02	.879+01	.393+00	.382+01
.102+02	.139+02	.387+00	.605+01
.104+02	.221+02	.344+00	.959+01
.106+02	.350+02	.280+00	.152+02
.108+02	.555+02	.212+00	.241+02
.110+02	.879+02	.151+00	.382+02

BUFFER CAPACITY TABLES

TEMP(DEGC)=25.0   IONIC STRENGTH= .0010
BUFFER CAPACITY=F+Y(ALK.-Z)
BUFFER CAPACITY IN PPM AS CACO3/PH
ALKALINITY IN PPM AS CACO3

BUFFER CAPACITY FACTORS:F,Y,Z

PH	F	Y	Z
.500+01	.119+01	.220+01	-.518+00
.520+01	.753+00	.215+01	-.327+00
.540+01	.475+00	.207+01	-.206+00
.560+01	.300+00	.196+01	-.130+00
.580+01	.190+00	.180+01	-.817-01
.600+01	.120+00	.159+01	-.513-01
.620+01	.771-01	.135+01	-.319-01
.640+01	.505-01	.109+01	-.193-01
.660+01	.347-01	.830+00	-.109-01
.680+01	.264-01	.605+00	-.494-02
.700+01	.238-01	.423+00	-.455-05
.720+01	.264-01	.287+00	.493-02
.740+01	.347-01	.192+00	.109-01
.760+01	.504-01	.128+00	.193-01
.780+01	.771-01	.865-01	.318-01
.800+01	.120+00	.623-01	.512-01
.820+01	.190+00	.506-01	.817-01
.840+01	.300+00	.492-01	.130+00
.860+01	.475+00	.576-01	.206+00
.880+01	.752+00	.766-01	.326+00
.900+01	.119+01	.108+00	.517+00
.920+01	.189+01	.153+00	.820+00
.940+01	.299+01	.213+00	.130+01
.960+01	.474+01	.280+00	.206+01
.980+01	.752+01	.344+00	.326+01
.100+02	.119+02	.386+00	.517+01
.102+02	.189+02	.393+00	.820+01
.104+02	.299+02	.362+00	.130+02
.106+02	.474+02	.303+00	.206+02
.108+02	.752+02	.234+00	.326+02
.110+02	.119+03	.170+00	.517+02

BUFFER CAPACITY TABLES

TEMP(DEGC)=25.0   IONIC STRENGTH= .0025
BUFFER CAPACITY=F+Y(ALK.-Z)
BUFFER CAPACITY IN PPM AS CACO3/PH
ALKALINITY IN PPM AS CACO3

BUFFER CAPACITY FACTORS:F,Y,Z

PH	F	Y	Z
.500+01	.122+01	.220+01	-.528+00
.520+01	.767+00	.215+01	-.333+00
.540+01	.484+00	.207+01	-.210+00
.560+01	.306+00	.195+01	-.132+00
.580+01	.193+00	.179+01	-.833-01
.600+01	.123+00	.158+01	-.523-01
.620+01	.786-01	.134+01	-.325-01
.640+01	.514-01	.108+01	-.197-01
.660+01	.354-01	.820+00	-.112-01
.680+01	.269-01	.596+00	-.504-02
.700+01	.243-01	.417+00	-.464-05
.720+01	.269-01	.283+00	.503-02
.740+01	.353-01	.189+00	.111-01
.760+01	.514-01	.126+00	.197-01
.780+01	.786-01	.855-01	.324-01
.800+01	.123+00	.620-01	.522-01
.820+01	.193+00	.511-01	.833-01
.840+01	.306+00	.505-01	.132+00
.860+01	.484+00	.599-01	.210+00
.880+01	.767+00	.801-01	.333+00
.900+01	.121+01	.113+00	.527+00
.920+01	.193+01	.160+00	.836+00
.940+01	.305+01	.221+00	.132+01
.960+01	.484+01	.289+00	.210+01
.980+01	.766+01	.351+00	.333+01
.100+02	.121+02	.389+00	.527+01
.102+02	.193+02	.391+00	.836+01
.104+02	.305+02	.355+00	.132+02
.106+02	.484+02	.294+00	.210+02
.108+02	.766+02	.225+00	.333+02
.110+02	.121+03	.163+00	.527+02

BUFFER CAPACITY TABLES

TEMP(DEGC)=25.0   IONIC STRENGTH= .0050
BUFFER CAPACITY=F+Y(ALK.-Z)
BUFFER CAPACITY IN PPM AS CACO3/PH
ALKALINITY IN PPM AS CACO3

BUFFER CAPACITY FACTORS:F,Y,Z

PH	F	Y	Z
.500+01	.124+01	.220+01	-.539+00
.520+01	.783+00	.215+01	-.340+00
.540+01	.494+00	.206+01	-.214+00
.560+01	.312+00	.194+01	-.135+00
.580+01	.197+00	.178+01	-.851-01
.600+01	.125+00	.157+01	-.533-01
.620+01	.803-01	.133+01	-.331-01
.640+01	.525-01	.106+01	-.201-01
.660+01	.361-01	.809+00	-.114-01
.680+01	.275-01	.587+00	-.514-02
.700+01	.248-01	.410+00	-.474-05
.720+01	.275-01	.278+00	.513-02
.740+01	.361-01	.185+00	.114-01
.760+01	.525-01	.124+00	.201-01
.780+01	.802-01	.845-01	.331-01
.800+01	.125+00	.618-01	.533-01
.820+01	.197+00	.516-01	.850-01
.840+01	.312+00	.519-01	.135+00
.860+01	.494+00	.624-01	.214+00
.880+01	.783+00	.841-01	.340+00
.900+01	.124+01	.119+00	.538+00
.920+01	.197+01	.168+00	.853+00
.940+01	.311+01	.230+00	.135+01
.960+01	.494+01	.298+00	.214+01
.980+01	.782+01	.358+00	.340+01
.100+02	.124+02	.392+00	.538+01
.102+02	.197+02	.388+00	.853+01
.104+02	.311+02	.348+00	.135+02
.106+02	.494+02	.285+00	.214+02
.108+02	.782+02	.216+00	.340+02
.110+02	.124+03	.155+00	.538+02

BUFFER CAPACITY TABLES

TEMP(DEGC)=25.0   IONIC STRENGTH= .0075
BUFFER CAPACITY=F+Y(ALK.-Z)
BUFFER CAPACITY IN PPM AS CACO3/PH
ALKALINITY IN PPM AS CACO3

BUFFER CAPACITY FACTORS:F,Y,Z

PH	F	Y	Z
.500+01	.126+01	.220+01	-.547+00
.520+01	.795+00	.214+01	-.345+00
.540+01	.502+00	.206+01	-.218+00
.560+01	.317+00	.194+01	-.137+00
.580+01	.200+00	.178+01	-.864-01
.600+01	.127+00	.157+01	-.542-01
.620+01	.815-01	.132+01	-.337-01
.640+01	.533-01	.106+01	-.204-01
.660+01	.367-01	.802+00	-.116-01
.680+01	.279-01	.581+00	-.522-02
.700+01	.252-01	.405+00	-.481-05
.720+01	.279-01	.274+00	.521-02
.740+01	.366-01	.183+00	.116-01
.760+01	.533-01	.122+00	.204-01
.780+01	.814-01	.837-01	.336-01
.800+01	.127+00	.617-01	.541-01
.820+01	.200+00	.521-01	.863-01
.840+01	.317+00	.531-01	.137+00
.860+01	.501+00	.644-01	.217+00
.880+01	.794+00	.871-01	.345+00
.900+01	.126+01	.123+00	.547+00
.920+01	.200+01	.173+00	.966+00
.940+01	.316+01	.236+00	.137+01
.960+01	.501+01	.304+00	.218+01
.980+01	.794+01	.363+00	.345+01
.100+02	.126+02	.394+00	.547+01
.102+02	.200+02	.386+00	.866+01
.104+02	.316+02	.343+00	.137+02
.106+02	.501+02	.278+00	.218+02
.108+02	.794+02	.210+00	.345+02
.110+02	.126+03	.150+00	.547+02

BUFFER CAPACITY TABLES

TEMP(DEGC)=25.0   IONIC STRENGTH= .0100
BUFFER CAPACITY=F+Y(ALK.-Z)
BUFFER CAPACITY IN PPM AS CACO3/PH
ALKALINITY IN PPM AS CACO3

BUFFER CAPACITY FACTORS:F,Y,Z

PH	F	Y	Z
.500+01	.128+01	.220+01	-.554+00
.520+01	.805+00	.214+01	-.349+00
.540+01	.508+00	.206+01	-.220+00
.560+01	.321+00	.194+01	-.139+00
.580+01	.203+00	.177+01	-.874-01
.600+01	.129+00	.156+01	-.548-01
.620+01	.825-01	.131+01	-.341-01
.640+01	.540-01	.105+01	-.207-01
.660+01	.371-01	.795+00	-.117-01
.680+01	.283-01	.575+00	-.529-02
.700+01	.255-01	.401+00	-.487-05
.720+01	.282-01	.271+00	.528-02
.740+01	.371-01	.181+00	.117-01
.760+01	.539-01	.121+00	.206-01
.780+01	.824-01	.832-01	.340-01
.800+01	.129+00	.616-01	.548-01
.820+01	.203+00	.525-01	.874-01
.840+01	.321+00	.541-01	.139+00
.860+01	.508+00	.661-01	.220+00
.880+01	.804+00	.897-01	.349+00
.900+01	.127+01	.127+00	.553+00
.920+01	.202+01	.178+00	.877+00
.940+01	.320+01	.242+00	.139+01
.960+01	.507+01	.310+00	.220+01
.980+01	.804+01	.366+00	.349+01
.100+02	.127+02	.394+00	.553+01
.102+02	.202+02	.384+00	.877+01
.104+02	.320+02	.338+00	.139+02
.106+02	.507+02	.273+00	.220+02
.108+02	.804+02	.205+00	.349+02
.110+02	.127+03	.145+00	.553+02

BUFFER CAPACITY TABLES

TEMP(DEGC)=25.0   IONIC STRENGTH= .0150
BUFFER CAPACITY=F+Y(ALK.-Z)
BUFFER CAPACITY IN PPM AS CACO3/PH
ALKALINITY IN PPM AS CACO3

BUFFER CAPACITY FACTORS:F,Y,Z

PH	F	Y	Z
.500+01	.130+01	.220+01	-.565+00
.520+01	.821+00	.214+01	-.356+00
.540+01	.518+00	.205+01	-.225+00
.560+01	.327+00	.193+01	-.142+00
.580+01	.207+00	.176+01	-.892-01
.600+01	.131+00	.155+01	-.559-01
.620+01	.842-01	.130+01	-.348-01
.640+01	.551-01	.104+01	-.211-01
.660+01	.379-01	.785+00	-.119-01
.680+01	.288-01	.567+00	-.539-02
.700+01	.260-01	.394+00	-.497-05
.720+01	.288-01	.267+00	.538-02
.740+01	.378-01	.178+00	.119-01
.760+01	.550-01	.119+00	.211-01
.780+01	.841-01	.823-01	.347-01
.800+01	.131+00	.616-01	.559-01
.820+01	.207+00	.533-01	.891-01
.840+01	.327+00	.557-01	.142+00
.860+01	.518+00	.689-01	.225+00
.880+01	.820+00	.940-01	.356+00
.900+01	.130+01	.133+00	.564+00
.920+01	.206+01	.186+00	.895+00
.940+01	.327+01	.251+00	.142+01
.960+01	.518+01	.318+00	.225+01
.980+01	.820+01	.372+00	.356+01
.100+02	.130+02	.395+00	.564+01
.102+02	.206+02	.380+00	.895+01
.104+02	.327+02	.330+00	.142+02
.106+02	.518+02	.264+00	.225+02
.108+02	.820+02	.196+00	.356+02
.110+02	.130+03	.139+00	.564+02

BUFFER CAPACITY TABLES

TEMP(DEGC)=90.0   IONIC STRENGTH= .0010
BUFFER CAPACITY=F+Y(ALK.-Z)
BUFFER CAPACITY IN PPM AS CACO3/PH
ALKALINITY IN PPM AS CACO3

BUFFER CAPACITY FACTORS:F,Y,Z

PH	F	Y	Z
.500+01	.120+01	.220+01	-.516+00
.520+01	.760+00	.214+01	-.324+00
.540+01	.486+00	.205+01	-.201+00
.560+01	.318+00	.193+01	-.122+00
.580+01	.218+00	.176+01	-.693-01
.600+01	.166+00	.154+01	-.316-01
.620+01	.149+00	.129+01	-.668-03
.640+01	.164+00	.103+01	.301-01
.660+01	.215+00	.778+00	.674-01
.680+01	.312+00	.562+00	.119+00
.700+01	.477+00	.390+00	.197+00
.720+01	.745+00	.264+00	.317+00
.740+01	.117+01	.177+00	.505+00
.760+01	.185+01	.119+00	.803+00
.780+01	.294+01	.834-01	.127+01
.800+01	.465+01	.642-01	.202+01
.820+01	.737+01	.578-01	.320+01
.840+01	.117+02	.629-01	.507+01
.860+01	.185+02	.798-01	.804+01
.880+01	.293+02	.110+00	.127+02
.900+01	.465+02	.154+00	.202+02
.920+01	.737+02	.213+00	.320+02
.940+01	.117+03	.280+00	.507+02
.960+01	.185+03	.344+00	.804+02
.980+01	.293+03	.386+00	.127+03
.100+02	.465+03	.393+00	.202+03
.102+02	.737+03	.362+00	.320+03
.104+02	.117+04	.303+00	.507+03
.106+02	.185+04	.235+00	.804+03
.108+02	.293+04	.171+00	.127+04
.110+02	.465+04	.118+00	.202+04

BUFFER CAPACITY TABLES

TEMP(DEGC)=90.0   IONIC STRENGTH= .0025
BUFFER CAPACITY=F+Y(ALK.-Z)
BUFFER CAPACITY IN PPM AS CACO3/PH
ALKALINITY IN PPM AS CACO3

BUFFER CAPACITY FACTORS:F,Y,Z

PH	F	Y(	Z
.500+01	.122+01	.219+01	-.526+00
.520+01	.775+00	.213+01	-.330+00
.540+01	.496+00	.205+01	-.205+00
.560+01	.324+00	.192+01	-.124+00
.580+01	.223+00	.175+01	-.707-01
.600+01	.169+00	.153+01	-.322-01
.620+01	.152+00	.128+01	-.681-03
.640+01	.167+00	.102+01	.307-01
.660+01	.219+00	.768+00	.687-01
.680+01	.318+00	.554+00	.122+00
.700+01	.486+00	.384+00	.201+00
.720+01	.759+00	.260+00	.323+00
.740+01	.120+01	.174+00	.515+00
.760+01	.189+01	.118+00	.818+00
.780+01	.299+01	.828-01	.130+01
.800+01	.474+01	.644-01	.206+01
.820+01	.751+01	.589-01	.326+01
.840+01	.119+02	.651-01	.517+01
.860+01	.189+02	.832-01	.819+01
.880+01	.299+02	.115+00	.130+02
.900+01	.474+02	.161+00	.206+02
.920+01	.751+02	.221+00	.326+02
.940+01	.119+03	.288+00	.517+02
.960+01	.189+03	.350+00	.820+02
.980+01	.299+03	.389+00	.130+03
.100+02	.474+03	.392+00	.206+03
.102+02	.751+03	.356+00	.326+03
.104+02	.119+04	.295+00	.517+03
.106+02	.189+04	.226+00	.820+03
.108+02	.299+04	.163+00	.130+04
.110+02	.474+04	.113+00	.206+04

BUFFER CAPACITY TABLES

TEMP(DEGC)=90.0   IONIC STRENGTH= .0050
BUFFER CAPACITY=F+Y(ALK.-Z)
BUFFER CAPACITY IN PPM AS CACO3/PH
ALKALINITY IN PPM AS CACO3

BUFFER CAPACITY FACTORS:F,Y,Z

PH	F	Y	Z
.500+01	.125+01	.219+01	-.537+00
.520+01	.791+00	.213+01	-.337+00
.540+01	.506+00	.204+01	-.209+00
.560+01	.331+00	.191+01	-.127+00
.580+01	.227+00	.174+01	-.721-01
.600+01	.172+00	.152+01	-.329-01
.620+01	.155+00	.127+01	-.695-03
.640+01	.171+00	.101+01	.313-01
.660+01	.224+00	.758+00	.701-01
.680+01	.325+00	.545+00	.124+00
.700+01	.496+00	.378+00	.205+00
.720+01	.775+00	.256+00	.330+00
.740+01	.122+01	.171+00	.526+00
.760+01	.193+01	.116+00	.835+00
.780+01	.306+01	.822-01	.133+01
.800+01	.484+01	.647-01	.210+01
.820+01	.767+01	.602-01	.333+01
.840+01	.122+02	.675-01	.528+01
.860+01	.193+02	.871-01	.837+01
.880+01	.305+02	.120+00	.133+02
.900+01	.484+02	.168+00	.210+02
.920+01	.767+02	.230+00	.333+02
.940+01	.122+03	.297+00	.528+02
.960+01	.193+03	.357+00	.837+02
.980+01	.305+03	.392+00	.133+03
.100+02	.484+03	.389+00	.210+03
.102+02	.767+03	.349+00	.333+03
.104+02	.122+04	.286+00	.528+03
.106+02	.193+04	.217+00	.837+03
.108+02	.305+04	.156+00	.133+04
.110+02	.484+04	.107+00	.210+04

BUFFER CAPACITY TABLES

TEMP(DEGC)=90.0   IONIC STRENGTH= .0075
BUFFER CAPACITY=F+Y(ALK.-Z)
BUFFER CAPACITY IN PPM AS CACO3/PH
ALKALINITY IN PPM AS CACO3

BUFFER CAPACITY FACTORS:F,Y,Z

PH	F	Y	Z
.500+01	.126+01	.219+01	-.545+00
.520+01	.803+00	.213+01	-.342+00
.540+01	.514+00	.204+01	-.212+00
.560+01	.336+00	.191+01	-.129+00
.580+01	.231+00	.173+01	-.732-01
.600+01	.175+00	.151+01	-.334-01
.620+01	.157+00	.126+01	-.706-03
.640+01	.174+00	.998+00	.318-01
.660+01	.227+00	.750+00	.712-01
.680+01	.330+00	.539+00	.126+00
.700+01	.504+00	.373+00	.208+00
.720+01	.787+00	.252+00	.335+00
.740+01	.124+01	.169+00	.534+00
.760+01	.196+01	.115+00	.848+00
.780+01	.310+01	.818-01	.135+01
.800+01	.491+01	.649-01	.213+01
.820+01	.779+01	.612-01	.338+01
.840+01	.123+02	.694-01	.536+01
.860+01	.196+02	.901-01	.849+01
.880+01	.310+02	.125+00	.135+02
.900+01	.491+02	.174+00	.213+02
.920+01	.779+02	.236+00	.338+02
.940+01	.123+03	.304+00	.536+02
.960+01	.196+03	.362+00	.849+02
.980+01	.310+03	.394+00	.135+03
.100+02	.491+03	.386+00	.213+03
.102+02	.779+03	.343+00	.338+03
.104+02	.123+04	.279+00	.536+03
.106+02	.196+04	.211+00	.849+03
.108+02	.310+04	.150+00	.135+04
.110+02	.491+04	.103+00	.213+04

BUFFER CAPACITY TABLES

TEMP(DEGC)=90.0   IONIC STRENGTH= .0100
BUFFER CAPACITY=F+Y(ALK.-Z)
BUFFER CAPACITY IN PPM AS CACO3/PH
ALKALINITY IN PPM AS CACO3

BUFFER CAPACITY FACTORS:F,Y,Z

PH	F	Y	Z
.500+01	.128+01	.219+01	-.552+00
.520+01	.813+00	.213+01	-.346+00
.540+01	.520+00	.203+01	-.215+00
.560+01	.340+00	.190+01	-.131+00
.580+01	.234+00	.173+01	-.742-01
.600+01	.177+00	.151+01	-.338-01
.620+01	.159+00	.126+01	-.715-03
.640+01	.176+00	.991+00	.322-01
.660+81	.230+00	.744+00	.721-01
.680+01	.334+00	.534+00	.128+00
.700+01	.510+00	.369+00	.210+00
.720+01	.796+00	.250+00	.339+00
.740+01	.125+01	.167+00	.540+00
.760+01	.198+01	.114+00	.859+00
.780+01	.314+01	.815-01	.136+01
.800+01	.498+01	.652-01	.216+01
.820+01	.788+01	.621-01	.342+01
.840+01	.125+02	.710-01	.543+01
.860+01	.198+02	.926-01	.860+01
.880+01	.314+02	.128+00	.136+02
.900+01	.497+02	.179+00	.216+02
.920+01	.788+02	.242+00	.342+02
.940+01	.125+03	.309+00	.543+02
.960+01	.198+03	.366+00	.860+02
.980+01	.314+03	.394+00	.136+03
.100+02	.497+03	.384+00	.216+03
.102+02	.788+03	.339+00	.342+03
.104+02	.125+04	.274+00	.543+03
.106+02	.198+04	.205+00	.860+03
.108+02	.314+04	.146+00	.136+04
.110+02	.497+04	.997-01	.216+04

BUFFER CAPACITY TABLES

TEMP(DEGC)=90.0   IONIC STRENGTH= .0150
BUFFER CAPACITY=F+Y(ALK.-Z)
BUFFER CAPACITY IN PPM AS CACO3/PH
ALKALINITY IN PPM AS CACO3

BUFFER CAPACITY FACTORS:F,Y,Z

PH	F	Y	Z
.500+01	.131+01	.219+01	-.563+00
.520+01	.8:9+00	.212+01	-.353+00
.540+01	.531+00	.203+01	-.219+00
.560+01	.347+00	.190+01	-.133+00
.580+01	.238+00	.172+01	-.756-01
.600+01	.181+00	.150+01	-.345-01
.620+01	.163+00	.124+01	-.729-03
.640+01	.179+00	.980+00	.328-01
.660+01	.235+00	.734+00	.735-01
.680+01	.341+00	.526+00	.130+00
.700+01	.520+00	.364+00	.215+00
.720+01	.812+00	.246+00	.346+00
.740+01	.128+01	.165+00	.551+00
.760+01	.202+01	.112+00	.876+00
.780+01	.320+01	.810-01	.139+0.1
.800+01	.508+01	.657-01	.220+01
.820+01	.804+01	.635-01	.349+01
.840+01	.127+02	.737-01	.553+01
.860+01	.202+02	.968-01	.877+01
.880+01	.320+02	.134+00	.139+02
.900+01	.507+02	.186+00	.220+02
.920+01	.804+02	.251+00	.349+02
.940+01	.127+03	.318+00	.553+02
.960+01	.202+03	.372+00	.877+02
.980+01	.320+03	.395+00	.139+03
.100+02	.507+03	.380+00	.220+03
.102+02	.804+03	.331+00	.349+03
.104+02	.127+04	.265+00	.553+03
.106+02	.202+04	.197+00	.877+03
.108+02	.320+04	.139+00	.139+04
.110+02	.507+04	.947-01	.220+04

Appendix B

ALKALINITY-ACIDITY-pH EQUILIBRIUM DIAGRAMS

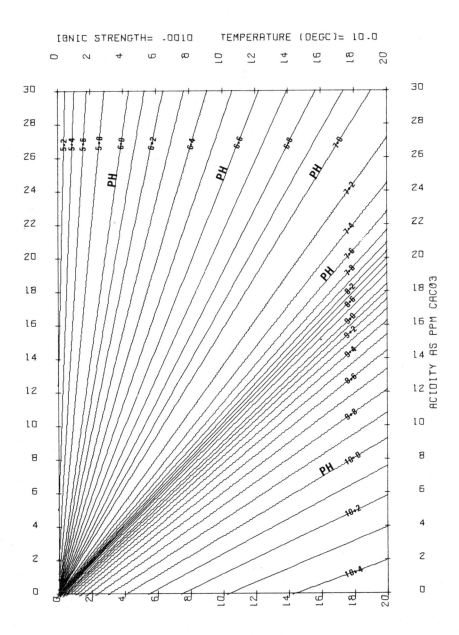

ALKALINITY AS PPM CACO3

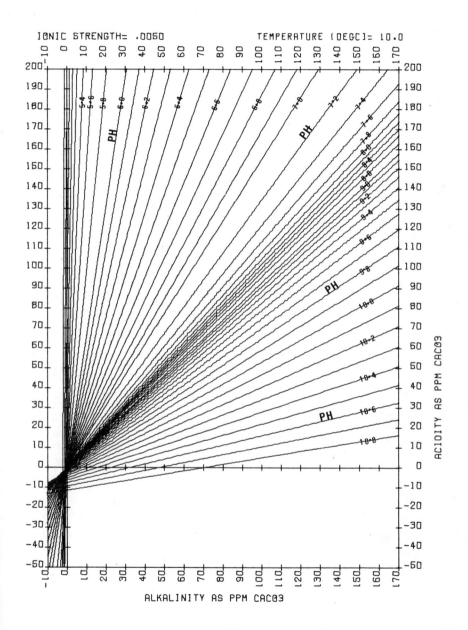

IONIC STRENGTH= .0060   TEMPERATURE (DEGC)= 10.0

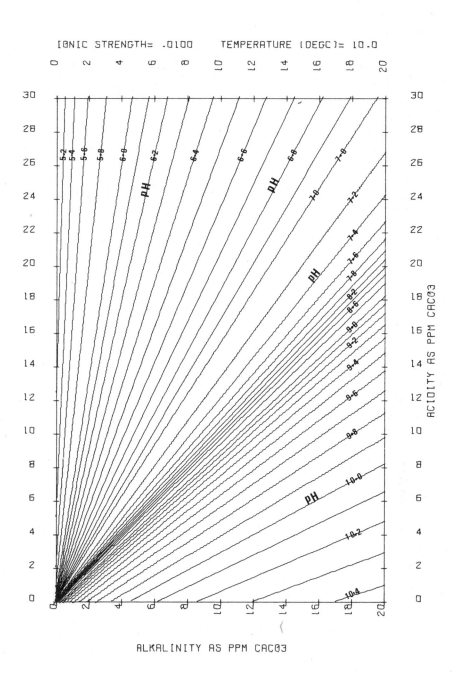

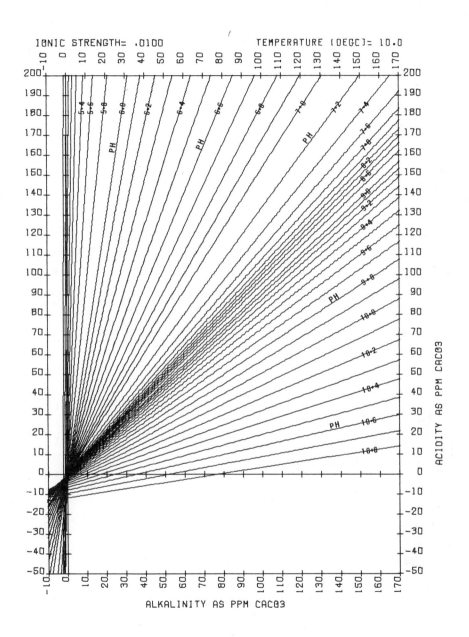

IONIC STRENGTH= .0010   TEMPERATURE (DEGC)= 15.0

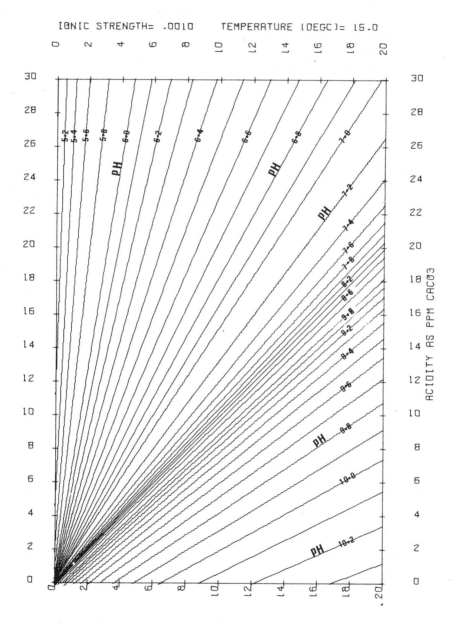

ACIDITY AS PPM CACO3

ALKALINITY AS PPM CACO3

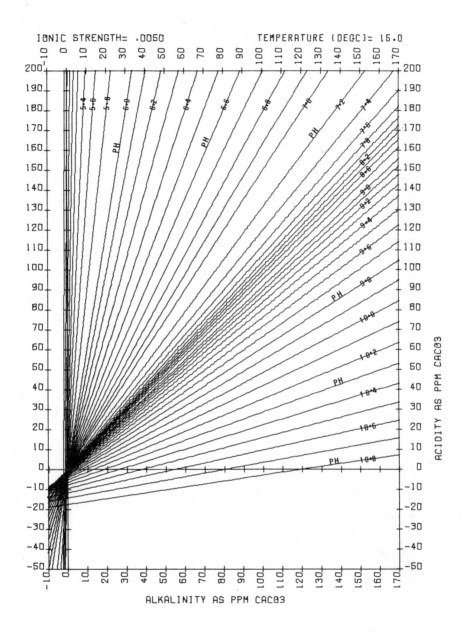

IONIC STRENGTH= .0100   TEMPERATURE (DEGC)= 15.0

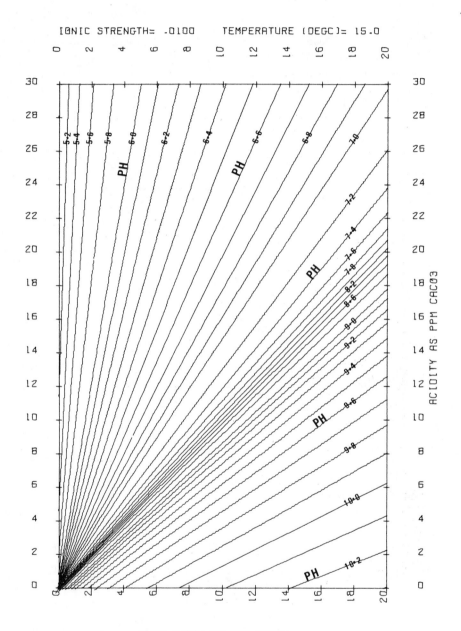

ALKALINITY AS PPM CACO3

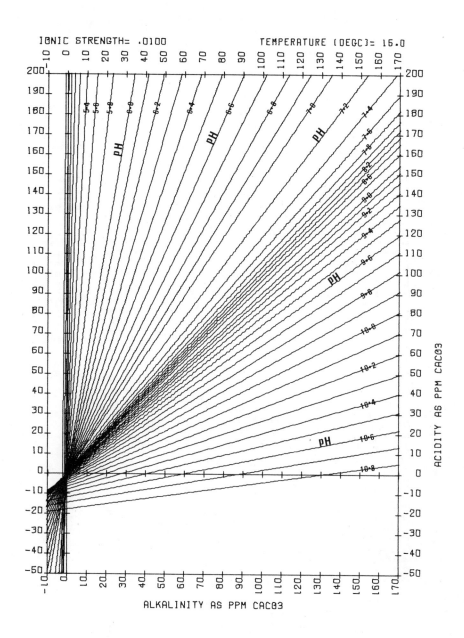

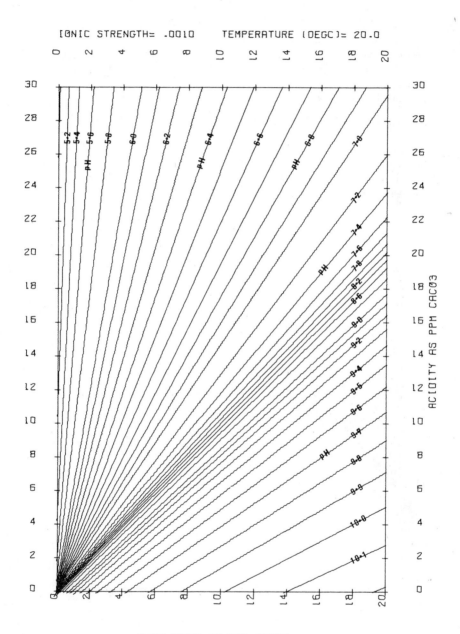

IONIC STRENGTH= .0010   TEMPERATURE (DEGC)= 20.0

ALKALINITY AS PPM CACO3

IONIC STRENGTH= .0050   TEMPERATURE [DEGC]= 20.0

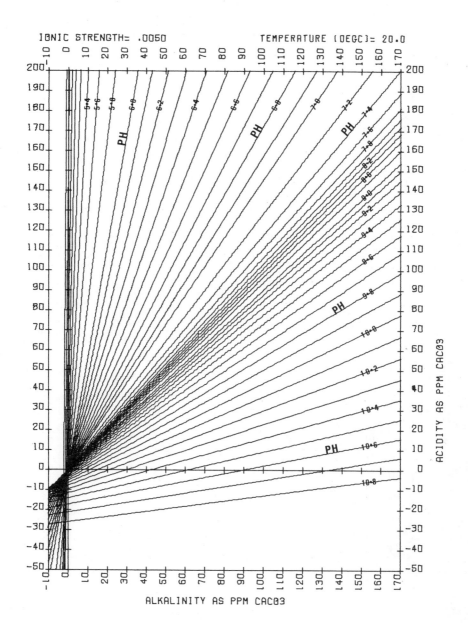

ALKALINITY AS PPM CACO3

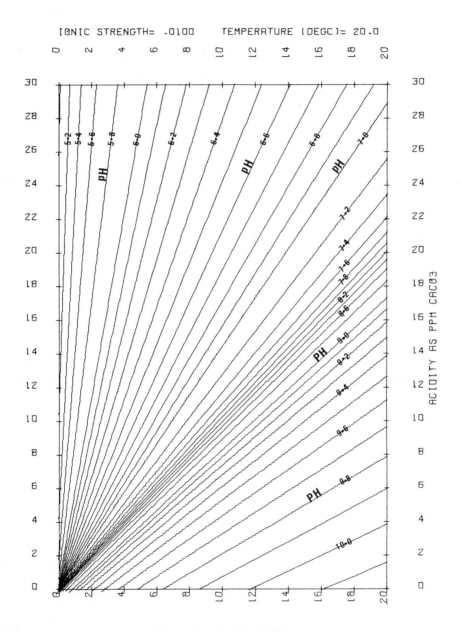

IONIC STRENGTH= .0100   TEMPERATURE (DEGC)= 20.0

ACIDITY AS PPM CACO3

ALKALINITY AS PPM CACO3

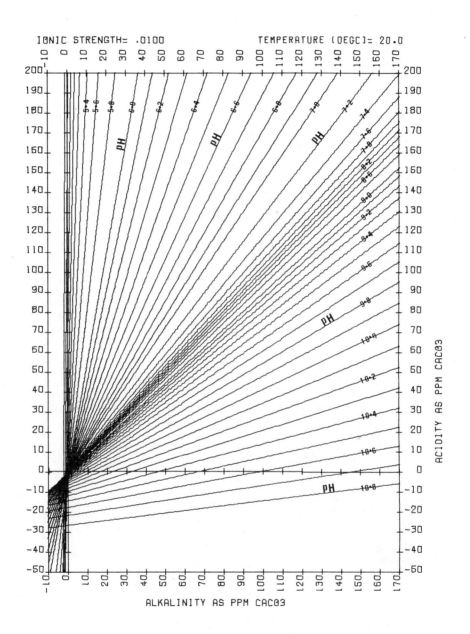

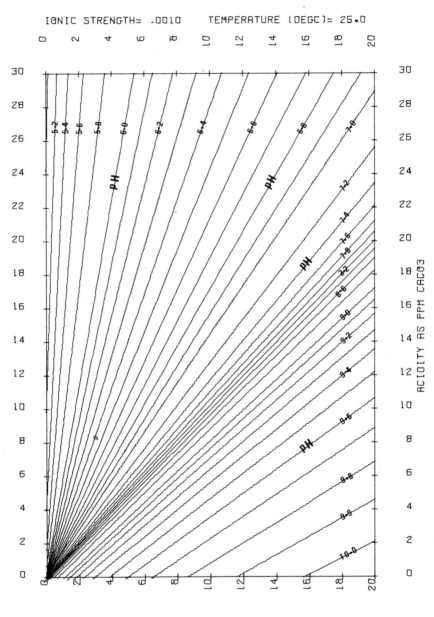

IONIC STRENGTH= .0010   TEMPERATURE (DEGC)= 25.0

ALKALINITY AS PPM CACO3

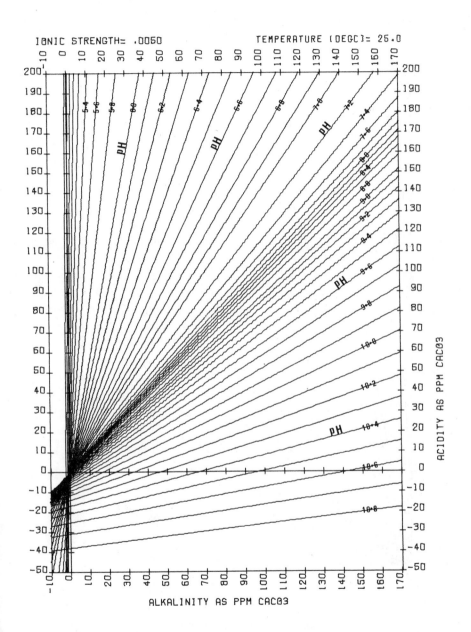

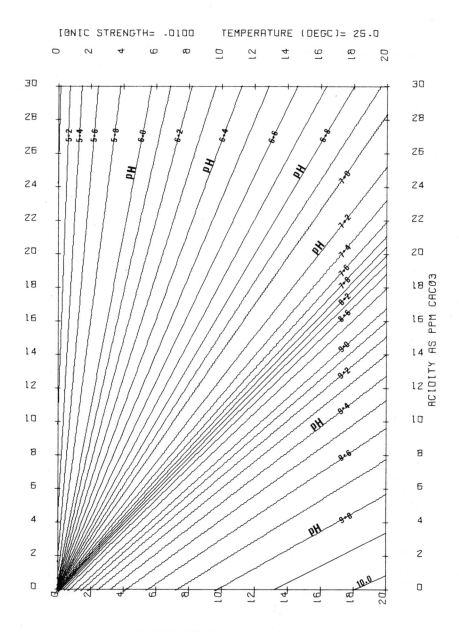

IONIC STRENGTH= .0100   TEMPERATURE (DEGC)= 25.0

ALKALINITY AS PPM CACO3

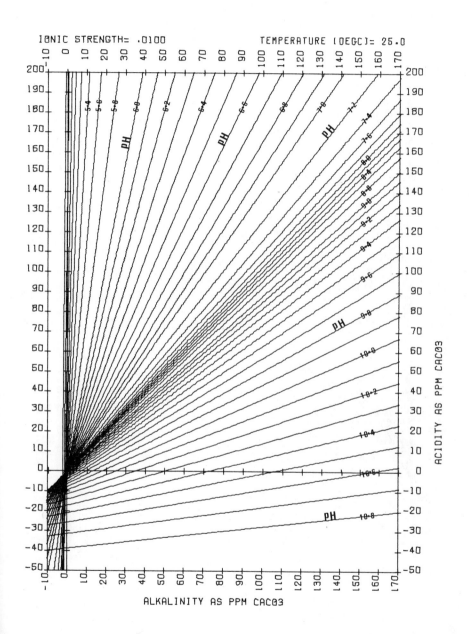

IONIC STRENGTH= .0100   TEMPERATURE (DEGC)= 25.0

ALKALINITY AS PPM CACO3

ACIDITY AS PPM CACO3

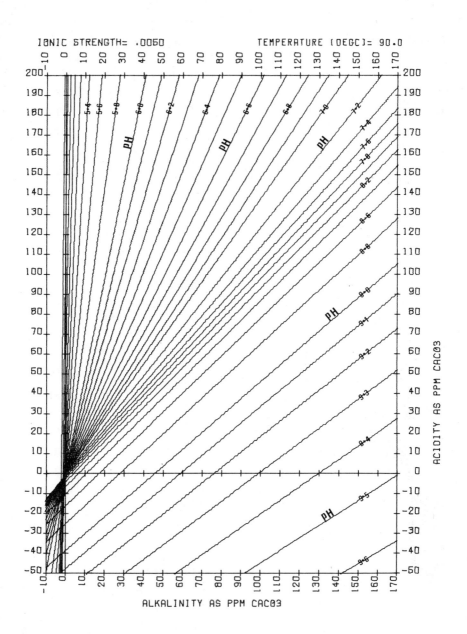

IONIC STRENGTH= .0060   TEMPERATURE (DEGC)= 90.0

ALKALINITY AS PPM CACO3

ACIDITY AS PPM CACO3

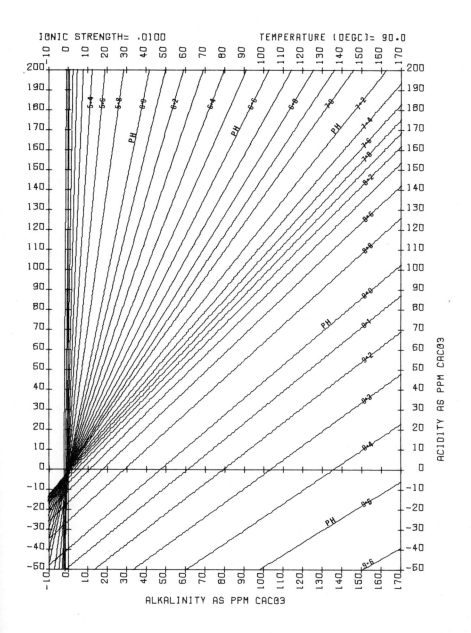

IONIC STRENGTH= .0100   TEMPERATURE (DEGC)= 90.0

ALKALINITY AS PPM CACO3

ACIDITY AS PPM CACO3

Appendix C

MODIFIED CALDWELL-LAWRENCE DIAGRAMS

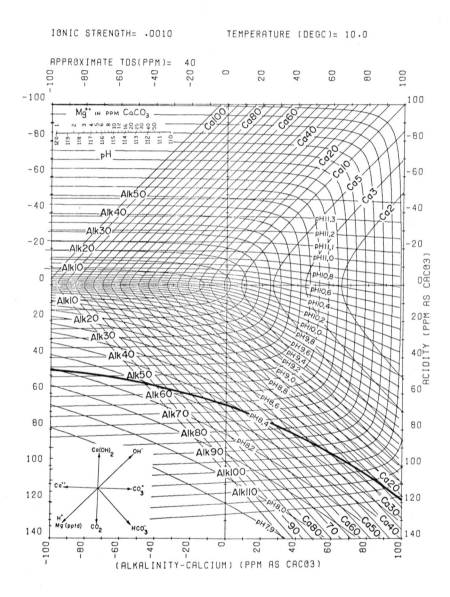

IONIC STRENGTH= .0010   TEMPERATURE (DEGC)= 10.0

APPROXIMATE TDS(PPM)=   40

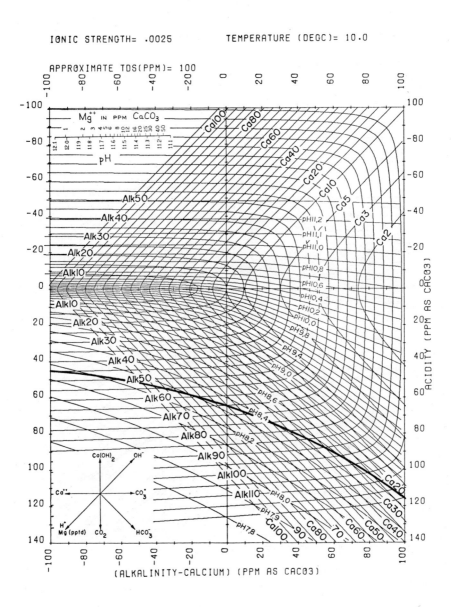

IONIC STRENGTH= .0025   TEMPERATURE (DEGC)= 10.0

APPROXIMATE TDS(PPM)= 100

Mg^{++} IN PPM CaCO$_3$

pH

Alk50
Alk40
Alk30
Alk20
Alk10
Alk10
Alk20
Alk30
Alk40
Alk50
Alk60
Alk70
Alk80
Alk90
Alk100
Alk110

Ca100 Ca80 Ca60 Ca40 Ca20 Ca10 Ca5 Ca3 Ca2

pH11.2
pH11.1
pH11.0
pH10.8
pH10.6
pH10.4
pH10.2
pH10.0
pH9.8
pH9.4
pH9.0
pH8.6
pH8.4
pH8.2
pH8.0
pH7.9
pH7.8

Ca20 Ca30 Ca40 Ca50 Ca60 70 80 90 Ca100

Ca(OH)$_2$ OH$^-$

Ca** CO$_3^-$

H$^+$
Mg (pptd)   CO$_2$ HCO$_3^-$

ACIDITY (PPM AS CACO3)

(ALKALINITY-CALCIUM) (PPM AS CACO3)

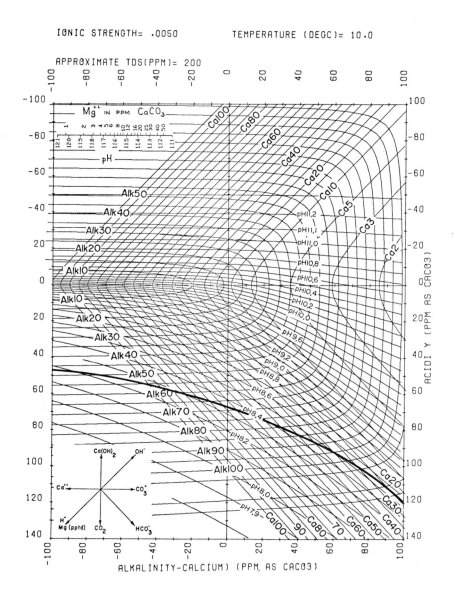

IONIC STRENGTH= .0050   TEMPERATURE (DEGC)= 10.0

APPROXIMATE TDS(PPM)= 200

ALKALINITY-CALCIUM) (PPM. AS CACO3)

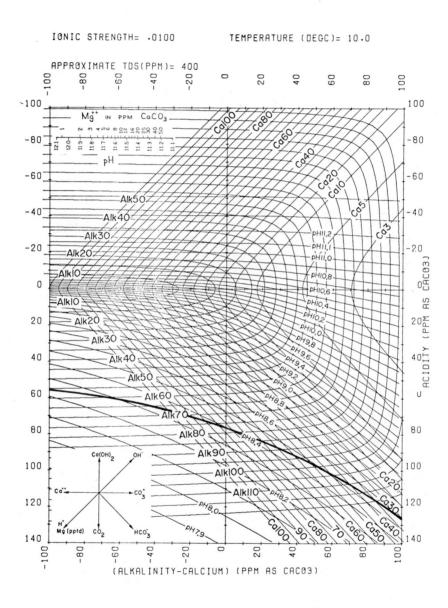

IONIC STRENGTH= .0100   TEMPERATURE (DEGC)= 10.0

APPROXIMATE TDS(PPM)= 400

(ALKALINITY-CALCIUM) (PPM AS CACO3)

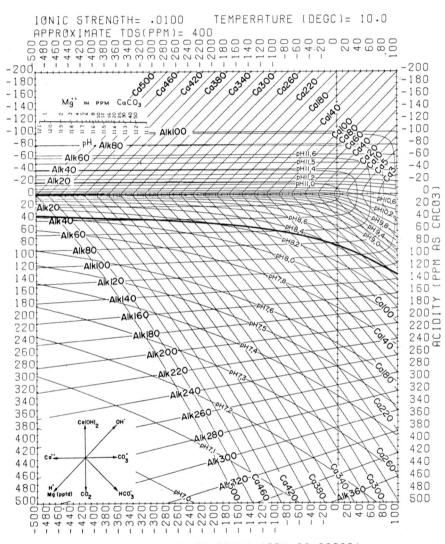

(ALKALINITY-CALCIUM) (PPM AS CACO3)

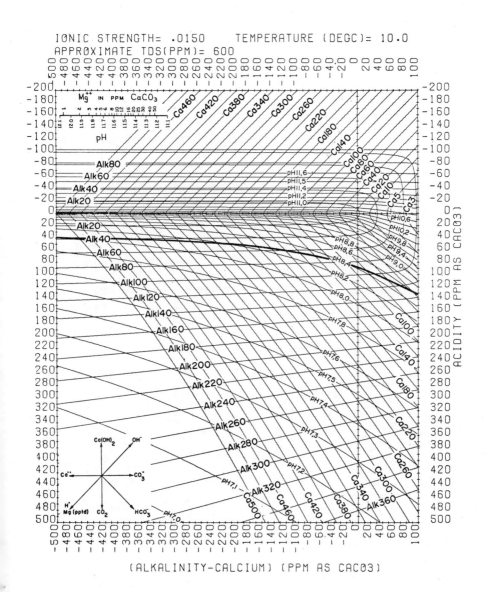

IONIC STRENGTH= .0150   TEMPERATURE (DEGC)= 10.0
APPROXIMATE TDS(PPM)= 600

(ALKALINITY-CALCIUM) (PPM AS CACO3)

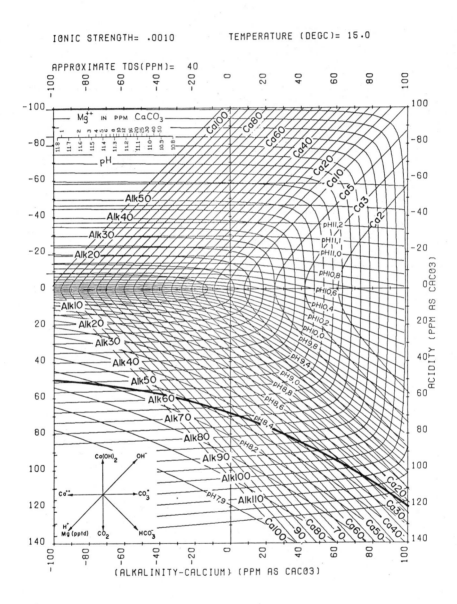

IONIC STRENGTH= .0010   TEMPERATURE (DEGC)= 15.0

APPROXIMATE TDS(PPM)=   40

(ALKALINITY-CALCIUM) (PPM AS CACO3)

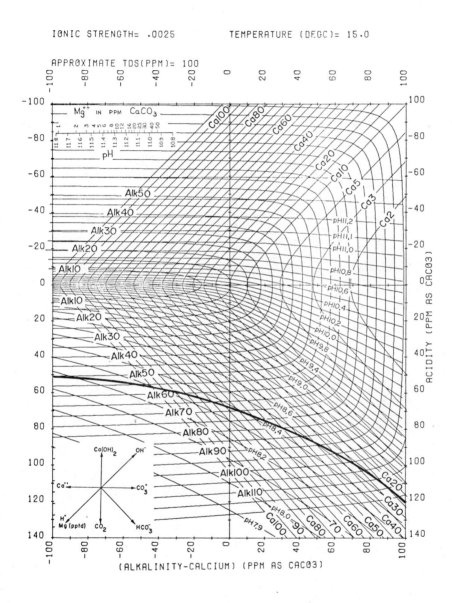

IONIC STRENGTH= .0025   TEMPERATURE (DEGC)= 15.0

APPROXIMATE TDS(PPM)= 100

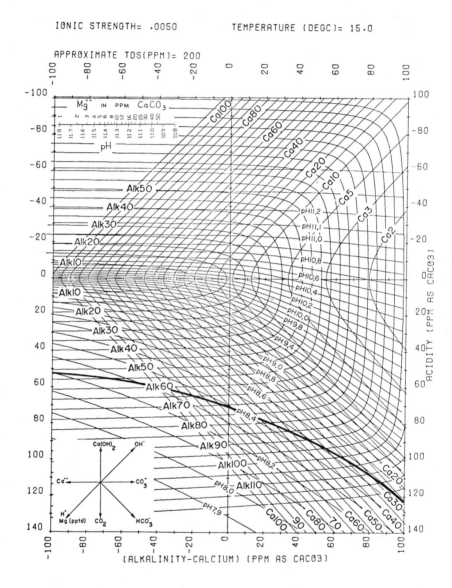

IONIC STRENGTH= .0050   TEMPERATURE (DEGC)= 15.0

APPROXIMATE TDS(PPM)= 200

(ALKALINITY-CALCIUM) (PPM AS CACO3)

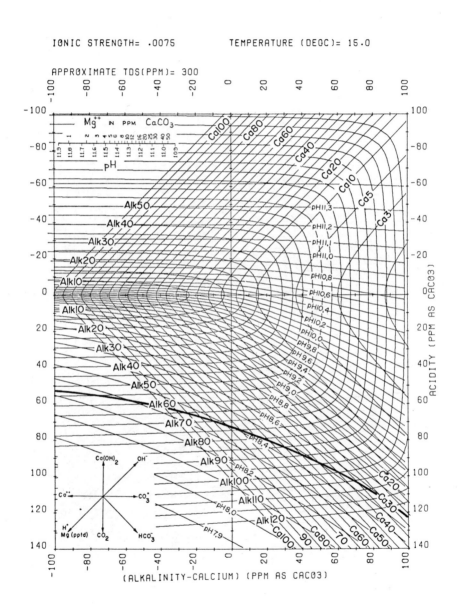

IONIC STRENGTH= .0075   TEMPERATURE (DEGC)= 15.0

APPROXIMATE TDS(PPM)= 300

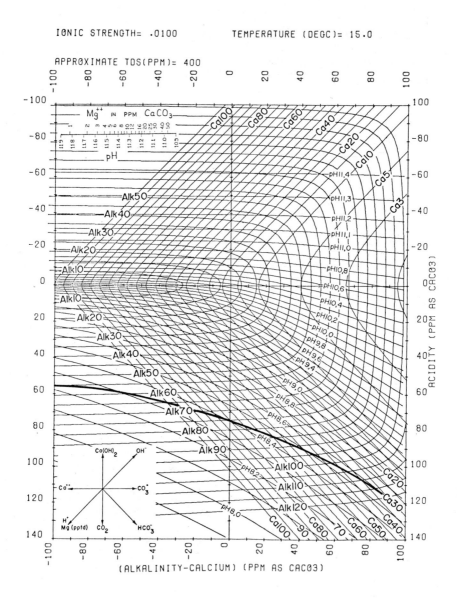

IONIC STRENGTH= .0100   TEMPERATURE (DEGC)= 15.0

APPROXIMATE TDS(PPM)= 400

(ALKALINITY-CALCIUM) (PPM AS CACO3)

ACIDITY (PPM AS CACO3)

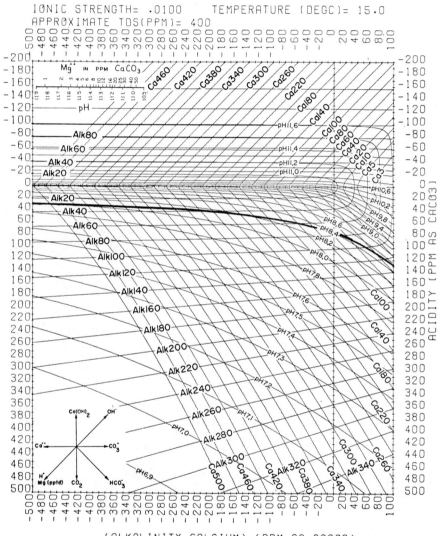

IONIC STRENGTH= .0100   TEMPERATURE (DEGC)= 15.0
APPROXIMATE TDS(PPM)= 400

(ALKALINITY-CALCIUM) (PPM AS CACO3)

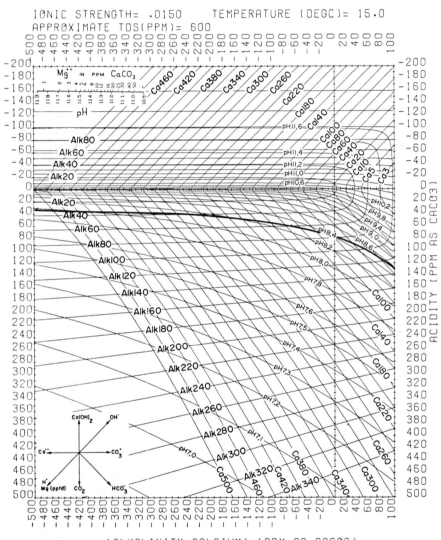

IONIC STRENGTH= .0150   TEMPERATURE (DEGC)= 15.0
APPROXIMATE TDS(PPM)= 600

(ALKALINITY-CALCIUM) (PPM AS CACO3)

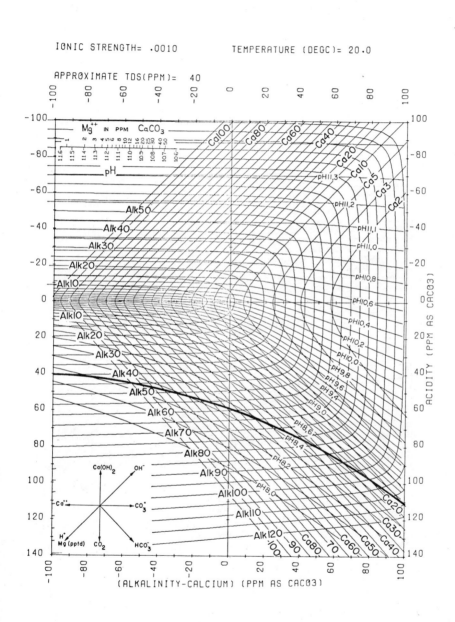

IONIC STRENGTH= .0010   TEMPERATURE (DEGC)= 20.0

APPROXIMATE TDS(PPM)=   40

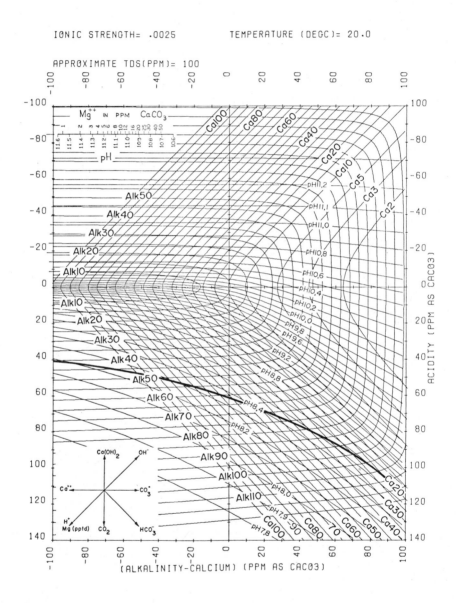

IONIC STRENGTH= .0025   TEMPERATURE (DEGC)= 20.0

APPROXIMATE TDS(PPM)= 100

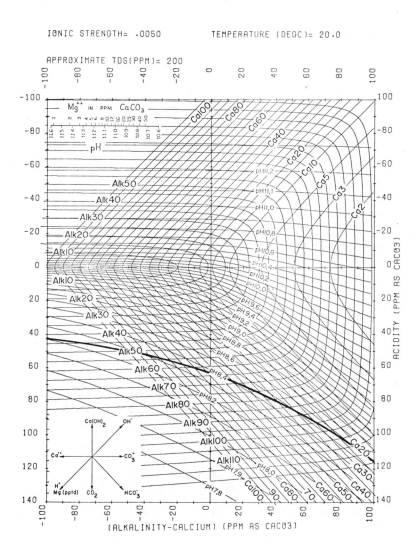

IONIC STRENGTH= .0050   TEMPERATURE (DEGC)= 20.0

APPROXIMATE TDS(PPM)= 200

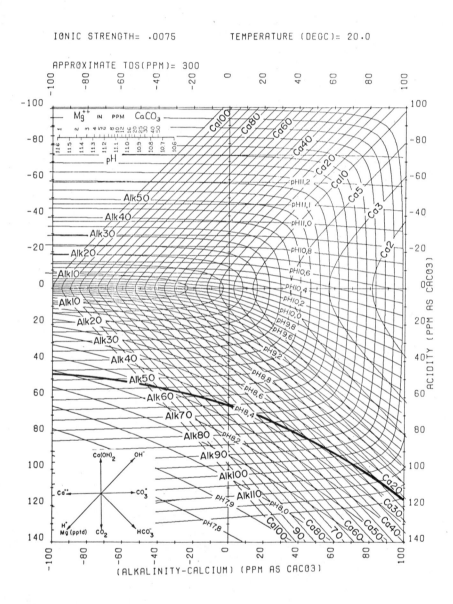

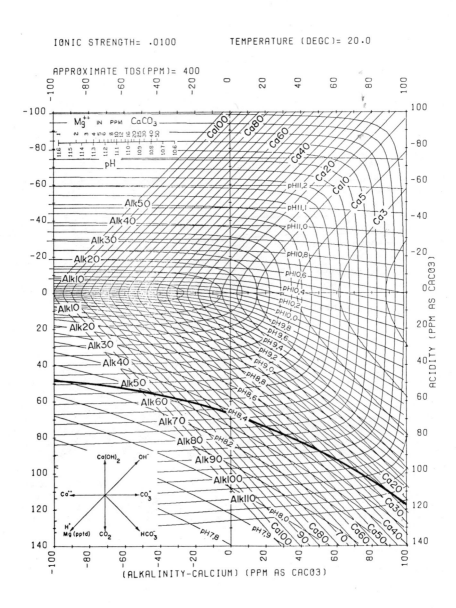

IONIC STRENGTH= .0100   TEMPERATURE (DEGC)= 20.0

APPROXIMATE TDS(PPM)= 400

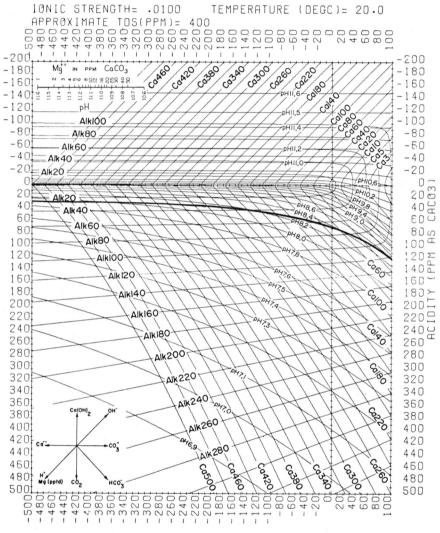

IONIC STRENGTH= .0100   TEMPERATURE (DEGC)= 20.0
APPROXIMATE TDS(PPM)= 400

(ALKALINITY-CALCIUM) (PPM AS CACO3)

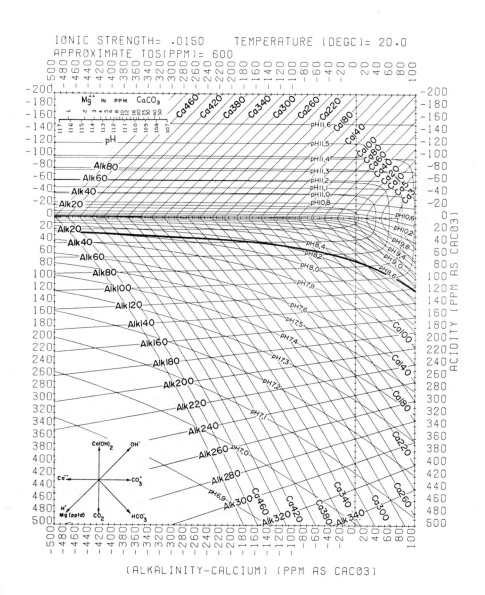

IONIC STRENGTH= .0150   TEMPERATURE (DEGC)= 20.0
APPROXIMATE TDS(PPM)= 600

(ALKALINITY-CALCIUM) (PPM AS CACO3)

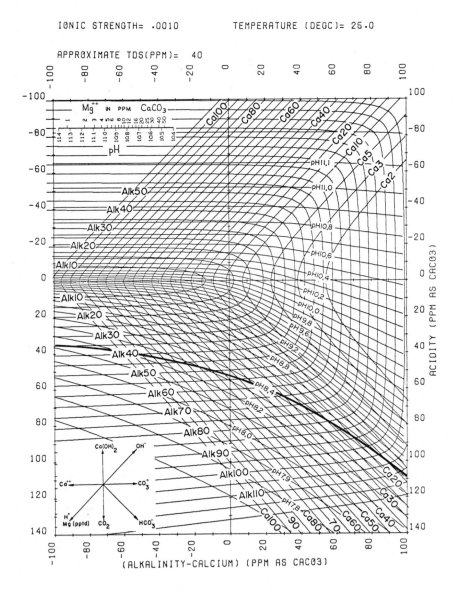

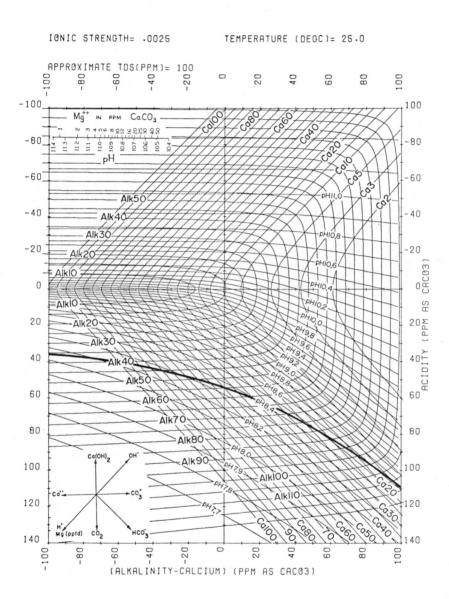

IONIC STRENGTH= .0025   TEMPERATURE (DEGC)= 25.0

APPROXIMATE TDS(PPM)= 100

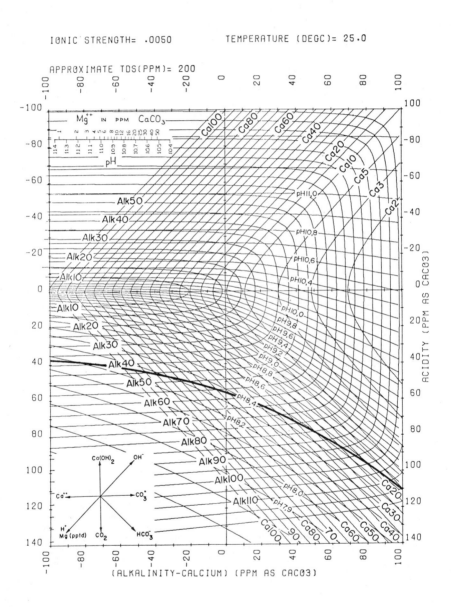

IONIC STRENGTH= .0050   TEMPERATURE (DEGC)= 25.0

APPROXIMATE TOS(PPM)= 200

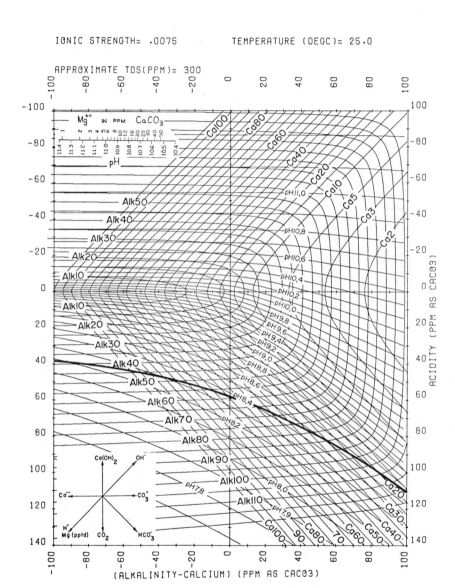

IONIC STRENGTH= .0075   TEMPERATURE (DEGC)= 25.0

APPROXIMATE TDS(PPM)= 300

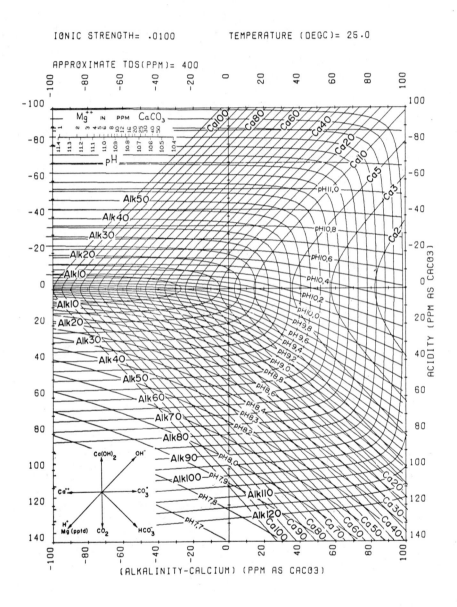

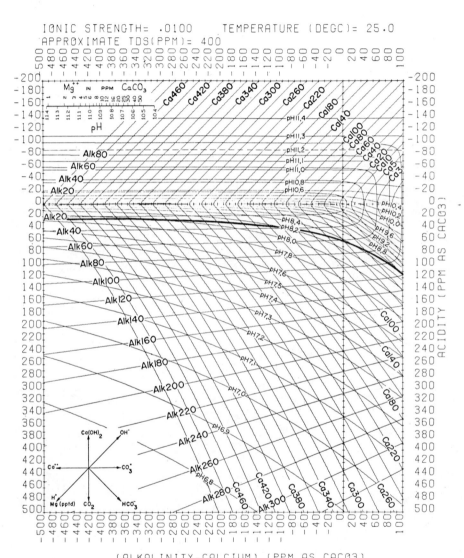

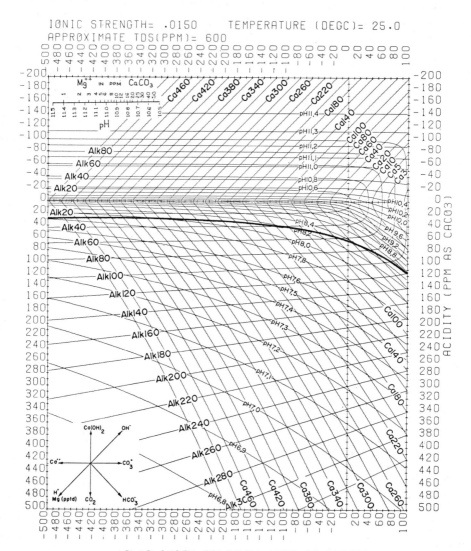

(ALKALINITY—CALCIUM) (PPM AS CACO3)

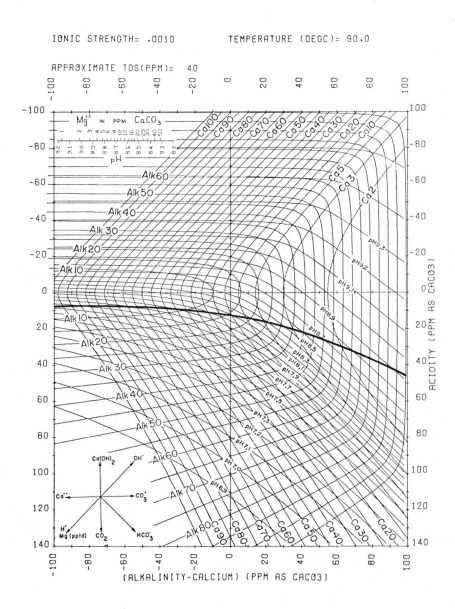

IONIC STRENGTH= .0010   TEMPERATURE (DEGC)= 90.0

APPROXIMATE TDS(PPM)=   40

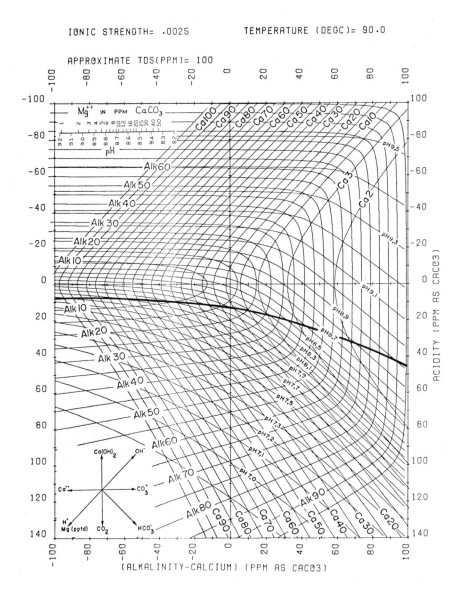

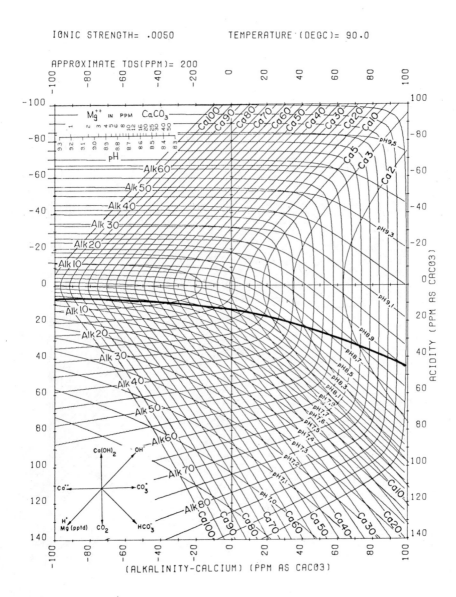

IONIC STRENGTH= .0050   TEMPERATURE (DEGC)= 90.0

APPROXIMATE TDS(PPM)= 200

(ALKALINITY-CALCIUM) (PPM AS CACO3)

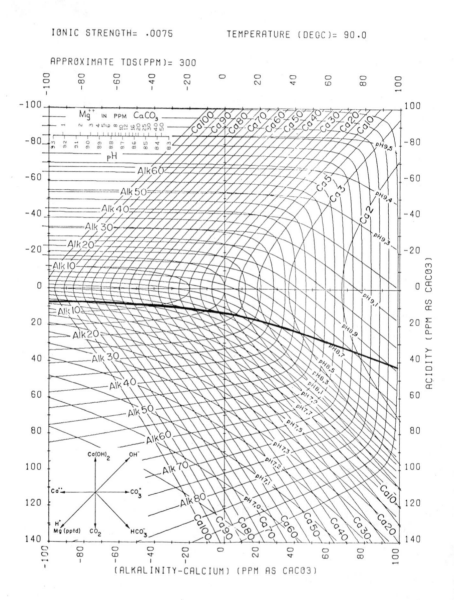

IONIC STRENGTH= .0075   TEMPERATURE (DEGC)= 90.0

APPROXIMATE TDS(PPM)= 300

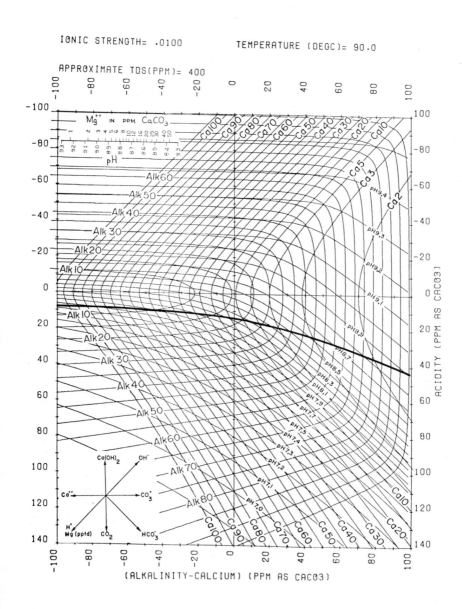

Appendix D

COMPUTER PROGRAM PRINTOUT

Three computer programs are listed in this Appendix. These programs are used together to plot a complete Modified Caldwell-
Lawrence Diagram as follows:

Program 1 plots lines representing selected pH, calcium and
Alkalinity values in a diagram with Acidity and (Alk-Ca) as cartesian coordinate parameters. The desired temperature and ionic
strength for which the diagram is to be drawn and the diagram
limits and scale are entered as input data.

Program 2 plots a nomogram giving pH values corresponding to
some fixed chosen Mg^{++} values for $Mg(OH)_2$ solubility. Temperature and ionic strength for which this nomogram is to be plotted
are entered as input data.

Program 3 plots a nomogram giving pH values corresponding to
chosen Alkalinity values for CO_2 equilibrium between air and
water. Ionic strength and temperature for which this nomogram is
to be plotted are entered as input data. Plotting the corresponding pH and Alkalinity values by hand in the diagram given by
Program 1 gives the 3 phase equilibrium line.

A more detailed description of each of these three programs is
given below.

Program 1.   This program plots Alkalinity, calcium and pH lines
in the Modified Caldwell-Lawrence Diagram. The diagram limits
and scale, the values of temperature and ionic strength and the
values of lines representing Alkalinity, calcium and pH to be
plotted are entered in the program as input data. This input
data is arranged as follows:

First Data Card

In this data card the desired 4 limits for the diagram (maximum
and minimum (Alk-Ca) values and the maximum and minimum Acidity
values) all as ppm $CaCO_3$ are entered as follows:

The first ten spaces of the data card are reserved for the maximum (Alk-Ca) value. (Format F 10.4)

Spaces 11 to 20 are reserved for the minimum (Alk-Ca) value. (Format F 10.4)

Spaces 21 to 30 are reserved for the minimum Acidity ordinate value. (Format F 10.4)

Spaces 31 to 40 are reserved for the maximum Acidity value. (Format F 10.4)

Second Data Card

In this data card the desired diagram scale is entered.   The scale is expressed as the number of inches in the diagram representing one ppm as $CaCO_3$, i.e. the number of inches representing either 1 ppm Acidity (as $CaCO_3$) for the vertical scale, or the number of inches representing 1 ppm (Alk-Ca) as $CaCO_3$ for the horizontal scale, e.g. 0,03.

This information is entered in the first 10 columns of this data card (Format F 10.5).

Third Data Card

In this data card the following three pieces of information are entered: the number of pH lines to be plotted; the number of Alkalinity lines to be plotted; the number of calcium lines to be plotted. This information is entered in the card as follows: Columns 1 and 2 are reserved for the integer number of pH lines to be plotted, e.g. 36 (Format 12); columns 3 and 4 are reserved for the integer number of Alkalinity lines to be plotted, e.g. 29 (Format 12); columns 5 and 6 are reserved for the integer number of calcium lines to be plotted (Format 12).

Fourth Data Card

In this data card the values of temperature (deg C) and ionic strength for which the diagram is to be plotted are recorded.

Columns 1 to 10 are reserved for the temperature value (Format F 10.4) and columns 11 to 20 are reserved for the ionic strength value.

Data Cards for the Values of pH Lines to be Plotted

After the Fourth Data Card a series of data cards are stacked each with one pH value per card. These pH values are the values of the pH lines to be plotted. The total number of cards, each with one pH value, must exactly equal that integer number entered in the Third Data Card giving the number of pH cards to be read. The pH value in each card must be entered in the first 10 columns (Format F 10.3).

Data Cards for the Values of Calcium Lines to be Plotted

After the data cards for the pH lines a series of data cards are stacked each with one calcium value per card. These calcium values are the values of the calcium lines to be plotted. The total number of cards, each with one calcium value, must exactly equal that integer number entered in the Third Data Card giving the number of calcium lines to be plotted. The cards containing these calcium values must be stacked in order of increasing calcium value, i.e. the card with the lowest calcium value must be the top card of the calcium data. The calcium value in each card must be entered in columns 1 to 10 (Format F 10.3).

Data Cards for the Values of Alkalinity Lines to be Plotted

After the data cards for the calcium lines a series of data cards are stacked each with one Alkalinity value per card. These Alkalinity values are the values of the Alkalinity lines to be plotted. The total number of cards, each containing one Alkalinity value, must exactly equal that integer number entered in the Third Data Card giving the number of Alkalinity lines to be plotted. The cards containing these Alkalinity values must be stacked in order of increasing Alkalinity value, i.e. the card with lowest

Alkalinity value must be the top card of the Alkalinity cards.
The Alkalinity value in each card must be entered in columns 1
to 10 (Format F 10.3).

Program 2. This program gives the corresponding pH and Mg^{++}
values for the solubility of $Mg(OH)_2$ in water at some fixed temperature and ionic strength. The arrangement of input data for
this program is set out below.

First Data Card

In this data card the integer value for the number of Mg^{++} values
for which pH is to be calculated is inserted. This integer value
is inserted in columns 1 and 2 (Format 12).

Second Data Card

In this data card are inserted the values of temperature (deg C)
and ionic strength at which solubility equilibrium of $Mg(OH)_2$ is
achieved. The value of temperature (deg C) is inserted in
columns 1 to 10 (Format F 10.4); the value of ionic strength in
columns 11 to 20 (Format F 10.4).

Data Cards for the Values of Mg^{++} for which the Corresponding pH
Values are to be Calculated for $Mg(OH)_2$ Solubility

After the Second Data Card a series of data cards are stacked;
each card contains one Mg^{++} values (expressed as ppm $CaCO_3$) for
which the corresponding pH is to be calculated. The Mg^{++} value
is inserted in columns 1 to 10 of each card (Format F 10.3). The
total number of cards, each containing one Mg^{++} value, must
exactly equal that integer value inserted in the First Data Card.

Program 3. This program gives corresponding pH and Alkalinity
values for equilibrium between CO_2 in the water at some fixed
temperature and ionic strength. The arrangement of input data
for this program is set out below.

First Data Card

The integer value of the number of pH values for which corresponding Alkalinity are to be calculated is inserted in this data
card. This integer value is inserted in columns 1 and 2 (Format
12).

Second Data Card

In this data card are inserted the values of temperature (deg C)
and ionic strength of the water at which CO_2 solubility equilibrium between water and air is achieved. The value of temperature
(deg C) is inserted in columns 1 to 10 (Format F 10.4); the value
of ionic strength is inserted in columns 11 to 20 (Format F 10.4).

Data Cards for the Values of pH for which Corresponding Alkalinities are to be Calculated for Air/Water CO_2 Equilibrium

After the Second Data Card a series of data cards are stacked
each card containing one pH value for which a corresponding
Alkalinity value is to be calculated. The pH value is in each
case inserted in columns 1 to 10 (Format F 10.3). The total
number of cards, each containing one pH value, should exactly
equal that integer value inserted in the First Data Card.

```
 1:C      ******************************************************
 2:C      CIVIL ENGINEERING DEPT.
 3:C      UNIVERSITY OF CAPE TOWN
 4:C      PROGRAM PLOTS THE MODIFIED
 5:C      CALDWELL-LAWRENCE DIAGRAM
 6:C      IBM PLOT SUBROUTINE IS USED
 7:C      ******************************************************
 8:       REAL K1,K2,KW,KS
 9:       DOUBLE PRECISION HCO3,FRACT
10:       DIMENSION C1(3000),C2(3000),PHI(60),STRING(5),AI(40),
11:      1CAI(40)
12:       M=5
13:       N=8
14:C      ******************************************************
15:C       FIRST DATA CARD:DIAGRAM LIMITS
16:       READ (N,31)YMAX,YMIN,XMAX,XMIN
17:  31   FORMAT(4F10.4)
18:C      WHERE:YMAX=NEGATIVE VALUE OF MIN(ALK-CA) ORDINATE
19:C            YMIN=NEGATIVE VALUE OF MAX(ALK-CA)ORDINATE
20:C             XMAX=MIN(ACIDITY)          (EG.-100.)
21:C             XMIN=MAX(ACIDITY)          (EG.140.)
22:C      ******************************************************
23:C      SECOND DATA CARD:DIAGRAM SCALE(SC)
24:       READ(N,32)SC
25:  32   FORMAT(F10.5)
26:C      SCALE IS NO. OF INCHES/PPM CACO3 (EG..03)
27:C      ******************************************************
28:C      THIRD DATA CARD:(1)NO. OF PH LINES=NX,(2)ALK. LINES=NZ,

29:C      (3)CA LINES=NY
30:       READ(N,33)NX,NZ,NY
31:  33   FORMAT(3I2)
32:C      ******************************************************
33:C      FOURTH DATA CARD:TEMP.(DEGC)=TS;IONIC STRENGTH=U
34:       READ(N,34)TS,U
35:  34   FORMAT(2F10.4)
36:C      ******************************************************
37:C      DATA CARDS FOR PH VALUES
38:       DO 35 J=1,NX
39:  35   READ(N,36)PHI(J)
40:  36   FORMAT(F10.3)
41:C      ******************************************************
42:C      DATA CARDS FOR CA VALUES
43:C      ONE CA VALUE PER CARD READ IN ASCENDING ORDER
44:       DO 37 J=1,NY
45:  37   READ(N,38)CAI(J)
46:  38   FORMAT(F10.3)
47:C      ******************************************************
```

```
48:C      DATA CARDS FOR ALK. VALUES
49:C      ONE ALK. VALUE/CARD;DATA TO BE READ IN ASCENDING ORDER
50:       DO 39 J=1,NZ
51:  39   READ(N,41)AI(J)
52:  41   FORMAT(F10.3)
53:C      ******************************************************
54:C      CALCULATE MONO+DI-VALENT ACTIVITY FACTORS
55:C      USING DAVIES EQUATION
56:       U1=U**0.5
57:       FD=2.*((U1/(1+U1))-0.2*U)
58:       FD=10.**FD
59:       FD=1./FD
60:       FM=0.5*((U1/(1+U1))-0.2*U)
61:       FM=10.**FM
62:       FM=1./FM
63:C      ******************************************************
64:C      CALCULATE EQUILIBRIUM CONSTANTS AT TEMP. TS
65:       TA=TS+273.
66:       PK1=(17052./TA)+215.21*ALOG(TA)/ALOG(10.)-(.12675*TA)
67:      1-545.56
68:       K1=1./10.**PK1
69:       PK2=(2902.39/TA)+(.02379*TA)-6.498
70:       K2=1./10.**PK2
71:       PKS=(.01183*TS)+8.03
72:       KS=1./10.**PKS
73:       PKW=(4787.3/TA)+(7.1321*ALOG(TA)/ALOG(10.))+(.01037*TA)
74:      1-22.801
75:       KW=1./10.**PKW
76:C      ******************************************************
77:C   CONVERT K VALUES TO ACCOUNT FOR ACTIVETY
78:       K1=K1/FM**2.
79:       K2=K2/FD
80:       KW=KW/FM**2.
81:       KS=KS/FD**2.
82:C      ******************************************************
83:C      CONVERT K VALUES TO PPM AS CACO3
84:       K1=K1*2.5*10.**4.
85:       K2=K2*10.**5.
86:       KW=KW*2.5*10.**9.
87:       KS=KS*10.**10.
88:       PI=3.14159
89:       SCS=-17./SC
90:C      ******************************************************
91:C      PLOT DIAGRAM OUTLINE
92:       CALL SCALF(SC,SC,SCS,SCS)
93:       NTMX=(XMIN-XMAX)/10.
94:       NTMY=(YMAX-YMIN)/10.
```

```
95:        CALL FGRID(0,-XMIN,YMIN,10.,NTMX)
96:        CALL FGRID(1,-XMAX,YMIN,10.,NTMY)
97:        CALL  FGRID(2,-XMAX,YMAX,10.,NTMX)
98:        CALL FGRID(3,-XMIN,YMAX,10.,NTMY)
99:        CALL FGRID(0,-XMIN,0.,10.,NTMX)
100:       CALL FGRID(1,0.,YMIN,10.,NTMY)
101:C      ********************************************************
102:C      LABEL THE DIAGRAM
103:C
104:       CALL PLTIME(30)
105:       FH=4.*SC
106:       YL=YMAX+FH*5./SC
107:       XB=-XMIN-FH*5./SC
108:       YR=YMIN
109:       XT=-XMAX
110:       NXX=-XMAX+XMIN+1
111:       DO 100 L=1,NXX,20
112:       NL=L+XMAX-1
113:       X=NL
114:       ENCODE(110,S)NL
115:       CALL PCHAR(-X,YR,FH,S,-PI/2,4)
116: 110   FORMAT(I4)
117: 100   CONTINUE
118:       DO 130 L=1,NXX,20
119:       NL=L+XMAX-1
120:       X=NL
121:       ENCODE(125,S)NL
122:       CALL PCHAR(-X,YL,FH,S,-PI/2,4)
123: 125   FORMAT(I4)
124: 130   CONTINUE
125:       NYY=YMAX-YMIN+1
126:       DO 150 L=1,NYY,20
127:       NL=L+YMIN-1
128:       Y=NL
129:       NL=-NL
130:       ENCODE(145,S)NL
131:       CALL PCHAR(XB,Y,FH,S,0.,4)
132: 145   FORMAT(I4)
133: 150   CONTINUE
134:       DO 160 L=1,NYY,20
135:       NL=L+YMIN-1
136:       Y=NL
137:       NL=-NL
138:       ENCODE(155,S)NL
139:       CALL PCHAR(XT,Y,FH,S,0.,4)
140: 155   FORMAT(I4)
141: 160   CONTINUE
142:C
143:       YR=YMIN-FH*7./SC
144:       YU=(YMAX+YMIN)/2.
```

```
145:        YL=YMAX+FH*6./SC
146:        YB=(YMAX+YMIN)/2.+18.*FH/SC
147:        XLR=(-XMIN-XMAX)/2.-11.*FH/SC
148:        XB=-XMIN-6.*FH/SC
149:        CALL PCHAR(XLR,YR,FH,'ACIDITY (PPM AS CACO3)',0.,22)
150:        CALL PCHAR(XB,YB,FH,'(ALKALINITY-CALCIUM) (PPM AS CACO3)

151:        .,-PI/2,35))
152:        CALL PCHAR(XLR,YL,FH,'ACIDITY (PPM AS CACO3)',0.,22)
153:        XUU=5.*FH/SC-XMAX
154:        XU=7.*FH/SC-XMAX
155:        ENCODE(182,STRING)U
156:        CALL PCHAR(XU,YMAX,FH,STRING,-PI/2,21)
157: 182    FORMAT('IONIC STRENGTH=',F6.4)
158:        ENCODE(183,STRING)TS
159:        CALL PCHAR(XU,YU,FH,STRING,-PI/2.,24)
160: 183    FORMAT('TEMPERATURE (DEGC)=',F5.1)
161:        TDS=(U*10.**5.)/2.5
162:        NTDS=TDS
163:        ENCODE(184,STRING)NTDS
164:        CALL PCHAR(XUU,YMAX,FH,STRING,-PI/2,25)
165: 184    FORMAT('APPROXIMATE TDS(PPM)=',I4)
166:C       *************************************************
167:C       CALCULATIONS TO PLOT PH LINES
168:        DO 90 J=1,NX
169:        PH=PHI(J)
170:        WRITE(M,2500)PH
171:2500    FORMAT(F10.4)
172:        H=1./(10.**PH)/FM
173:C   CONVERT H TO PPM CACO3
174:        H=(5.*10.**4.)*H
175:        OH=KW/H
176:        C2(1)=-YMAX
177:        JJ=1
178:        II=1
179:        DO 65 I=1,500
180:        B=C2(I)-OH+H
181:        B2=B**2.
182:        AC4=4.*((K2/H)+1.)*KS*H/K2
183:        Z=(B2+AC4)
184:        SRZ=Z**0.5
185:        A2=2.*((K2/H)+1.)
186:        HCO3=(B+SRZ)/A2
187:        CO2=HCO3*H/K1
188:        C1(I)=OH-HCO3-CO2-H
189:        C2(I+1)=C2(I)+2.
190:        IF(C1(I)+XMAX)50,50,49
```

```
191:   49   JJ=JJ+1
192:   50   IF(C1(I)+XMIN)70,55,55
193:   55   IF(C2(I)+YMIN)60,70,70
194:   60   II=II+1
195:   65   CONTINUE
196:C      ****************************************************
197:C      PLOT PH LINES
198:   70   CALL FPLOT(3,C1(JJ),-C2(JJ))
199:        II=II-1
200:        DO 80 I=JJ,II
201:        CALL FPLOT(2,C1(I),-C2(I))
202:   80   CONTINUE
203:   90    CONTINUE
204:        CALL FPLOT(3,0.,0.)
205:C      ****************************************************
206:C      CALCULATIONS TO PLOT ALKALINITY LINES
207:C      LL=0 LOWER HORIZONTAL ARM
208:C      LL=1 VERTICAL ARM
209:C      LL=2 UPPER HORIZONTAL ARM
210:        C2(1)=-YMAX
211:        LUC=1
212:        LK=1
213:        LL = 0
214:        DO 420 J=1,NZ
215:        A=AI(J)
216:        JJ=1
217:        WRITE(M,2510)A
218:2510   FORMAT(F10.4)
219:        II=1
220:        IF(LUC-1)212,212,211
221:  211   C1(1)=-XMIN
222:        LL=1
223:C      ****************************************************
224:C      CALCULATIONS FOR LOWER HORIZONTAL
225:C      LIMB OF ALKALINITY LINE
226:  212   DO 400 I=1,3000
227:        JJ=JJ+1
228:        IF(LL-1)220,260,330
229:  220   B= A - KS/(A-C2(I)+10.**(-10.))
230:        B2 = B**2.
231:        AC4 = (4.*KS*KW)/(K2*(A-C2(I)+10.**(-10.)))
232:        SRZ1=B2-AC4
233:        IF(SRZ1)2202,2202,2201
234:2202   C1(I)=C1(I-1)
235:        GO TO 260
236:2201   SRZ=SRZ1**0.5
237:        A2 =2*(KS/(K2*(A-C2(I)+10.**(-10.)))-1.)
238:        H = (B+SRZ)/A2
239:        HCO3 = KS*H/(K2*(A-C2(I)))
240:        CO2 = (H*HCO3)/K1
```

```
241:        C1(I) = KW/H-CO2-HCO3-H
242:        IF (C1(I)+XMIN)221,221,230
243: 221    LUC=2
244:        C1(1)=-XMIN
245:        GO TO 260
246: 230    IF(II-1)231,231,232
247: 231    X=C1(I)
248:        Z=C2(I)
249: 232    IF(C2(I)+YMIN)240,410,410
250: 240    IF(II-1)250,250,245
251: 245    II=II+1
252:        IF(C1(I)-C1(I-1)-0.6)250,260,260
253: 250    C2(I+1)=C2(I)+0.61
254:        II=II+1
255:        GO TO 400
256:C       ****************************************************
257:C       CALCULATIONS FOR VERTICAL
258:C       LIMB OF ALKALINITY LINE
259: 260    LL=1
260:        HCO3=1.
261: 259    Y=C1(I)-KW*(A-HCO3)/(K2*HCO3+KW)+(HCO3/K1)*(K2*HCO3+KW)/

262:        .(A-HCO3+10.**(-10.))+HCO3+(K2*HCO3+KW)/(A-HCO3)
263:        IF(Y)270,310,265
264: 265    HCO3=0.1*HCO3
265:        GO TO 259
266: 270    K=1
267:        FRACT=HCO3
268: 280    HCO3=HCO3+FRACT
269:        Y=C1(I)-KW*(A-HCO3)/(K2*HCO3+KW)+(HCO3/K1)*(K2*HCO3+KW)/

270:        .(A-HCO3+10.**(-10.))+HCO3+(K2*HCO3+KW)/(A-HCO3)
271:        IF(Y)280,310,290
272: 290    HCO3=HCO3-FRACT
273:        FRACT=FRACT*0.1
274:        IF(9-K)310,310,300
275: 300    K=K+1
276:        GO TO 280
277: 310    CONTINUE
278:        H=(K2*HCO3+KW)/(A-HCO3)
279:        C2(I)=A  -KS*H/(K2*HCO3)
280:        IF(II-1)315,315,317
281: 315    II=II+1
282:        X=C1(I)
283:        Z=C2(I)
284:        GO TO 320
285: 317    IF(C2(I-1)-C2(I)-0.6)320,330,330
286: 320    CONTINUE
287:        C1(I+1)=C1(I)+0.61
288:        IF(C1(I)+XMAX)325,325,410
```

```
289: 325   IF(C2(I)+YMIN)400,410,410
290:C      *********************************************
291:C      CALCULATIONS FOR UPPER HORIZONTAL
292:C      LIMB OF ALKALINITY CURVE
293: 330   LL=2
294:       B= A - KS/(A-C2(I)+10.**(-10.))
295:       B2 = B**2.
296:       AC4 = (4.*KS*KW)/(K2*(A-C2(I)+10.**(-10.)))
297:       SRZ = (B2-AC4)**0.5
298:       A2 =2*(KS/(K2*(A-C2(I)+10.**(-10.)))-1.)
299:       H=(B-SRZ)/A2
300:       HCO3 = KS*H/(K2*(A-C2(I)))
301:       CO2 = (H*HCO3)/K1
302:       C1(I) = KW/H-CO2-HCO3-H
303:       C2(I+1)=C2(I)-5.
304:       IF(C2(I)+YMAX)410,400,400
305: 400   CONTINUE
306:C      *********************************************
307:C      PLOT ALKALINITY CURVE
308: 410   CALL FPLOT(3,X,-Z)
309:       JJ=JJ-1
310:       DO 415 I=1,JJ
311:       CALL FPLOT(2,C1(I),-C2(I))
312: 415   CONTINUE
313:       LL=0
314: 420   CONTINUE
315:       CALL FPLOT(3,0.,0.)
316:C      *********************************************
317:C      CALCULATIONS FOR CALCIUM LINES
318:       II=1
319:       JJ=1
320:       DO 590 J=1,NY
321:       CA=CAI(J)
322:       WRITE(M,2520)CA
323:2520   FORMAT(F10.4)
324:       C1(1)=-XMAX
325:       DO 581 I=1,3000
326:       HCO3=10.**3.
327: 509   Y=C1(I)+(K2*CA*HCO3**2.)/(KS*K1)+HCO3-(KW*KS)/(HCO3*K2*
328:       1CA)
329:       .+(K2*CA*HCO3)/KS
330:       IF(Y)515,550,510
331: 510   HCO3=0.1*HCO3
332:       GO TO 509
333: 515   KK=1
334:       FRACT=HCO3
```

```
335:  520   HCO3=HCO3+FRACT
336:        Y=C1(I)+(K2*CA*HCO3**2.)/(KS*K1)+HCO3-(KW*KS)/
337:        1(HCO3*K2*CA)
338:        .+(K2*CA*HCO3)/KS
339:        IF(Y)520,520,530
340:  530    HCO3=HCO3-FRACT
341:        FRACT=FRACT*0.1
342:        IF(5-KK)550,540,540
343:  540   KK=KK+1
344:        GO TO 520
345:  550   CONTINUE
346:        CO3=KS/CA
347:        OH=(KW*KS)/(HCO3*K2*CA)
348:        H=KW/OH
349:        C2(I)=HCO3+CO3+OH-CA-H
350:        IF(C2(I)+YMAX)585,551,551
351:  551   IF(C2(I)+YMIN)560,555,555
352:  555   NABZ=-XMAX
353:        IF(II-NABZ)556,556,585
354:  556   JJ=JJ+1
355:  560   IF(C1(I)+XMIN)585,585,570
356:  570   C1(I+1)=C1(I)-1.
357:        II=II+1
358:  581   CONTINUE
359:C     ****************************************************
360:C     PLOT CALCIUM LINE
361:  585   CALL FPLOT(3,C1(JJ),-C2(JJ))
362:        DO 583 I=JJ,II
363:        CALL FPLOT(2,C1(I),-C2(I))
364:  583   CONTINUE
365:        II=1
366:        JJ=1
367:  590   CONTINUE
368:        CALL FPLOT(3,0.,0.)
369:        CALL PEND(200.,-YMAX)
370:        STOP
371:        END
```

```
 1:C     ************************************************************
 2:C     THIS PROGRAM CALCULATES PH FOR SELECTED MG++
 3:C     VALUES FOR  MG(OH)2 SOLUBILITY
 4:C     ************************************************************
 5:      REAL MG,KM,KW
 6:      DIMENSION TTS(50),UU(50),XMG(50)
 7:      M=5
 8:      N=8
 9:C     DATA INPUT
10:C     ************************************************************
11:C     READ IN 4 TEMPERATURES
12:C
13:      DO 700 J=1,4
14:      READ(N,701)TTS(J)
15: 701  FORMAT( )
16: 700  CONTINUE
17:C     ************************************************************
18:C     READ IN 7 IONIC STRENGTHS
19:C
20:      DO 710 NN=1,7
21:      READ(N,709)UU(NN)
22: 709  FORMAT( )
23: 710  CONTINUE
24:C     ************************************************************
25:C     THIRD DATA:DATA CARDS FOR MG++ VALUES(XMG)
26:C
27:      READ(N,701)(XMG(I),I=1,20)
28:      DO 790 J=1,4
29:      TA=TTS(J)+273.
30:      TS=TTS(J)
31:      DO 780 NN=1,7
32:      U=UU(NN)
33:C     ************************************************************
34:C     CALCULATE EQUILIBRIUM CONSTANTS  FOR INPUT TEMP.
35:C
36:      PKM=0.0175*TS+9.97
37:      KM=1./10.**PKM
38:      PKW=(4787.3/TA)+(7.1321*ALOG(TA)/ALOG(10.))+(.01037*TA)
39:      1-22.801
40:      KW=1./10.**PKW
41:C     ************************************************************
42:C     CALCULATE MONO AND DI VALENT ACTIVETY FACTORS
43:C
44:      U1=U**0.5
45:      FD=2.*((U1/(1+U1))-0.2*U)
46:      FD=10.**FD
47:      FD=1./FD
48:      FM=0.5*((U1/(1+U1))-0.2*U)
```

```
49:          FM=10.**FM
50:          FM=1./FM
51:C         ********************************************************
52:C         ADJUST EQ. CONSTANTS FOR IONIC STRENGTH
53:C
54:          KW=KW/FM**2.
55:          KM=KM/(FD*FM**2.)
56:C         ********************************************************
57:C         ADJUST CONSTANTS   TO PPM AS CACO3
58:C
59:          KW=KW*2.5*10.**9.
60:          KM=KM*2.5*10.**14.
61:C
62:C         ********************************************************
63:C         CALCULATE PH FOR INPUT MG++ VALUE
64:C         ********************************************************
65:          DO 760 I=1,20
66:          MG=XMG(I)
67:          H=KW*(MG/KM)**0.5
68:          HE=H*FM
69:          HE=HE/(5.*10.**4.)
70:          PHM=(ALOG(1./HE))/ALOG(10.)
71:          WRITE(M,750)MG,PHM,TTS(J),U
72:  750     FORMAT(4F10.6)
73:  760     CONTINUE
74:  780     CONTINUE
75:  790     CONTINUE
76:          STOP
77:          END
```

```
 1:C      THIS PROGRAM CALCULATES ALKALINITY VALUES CORRESPONDING

 2:C      TO INPUT  PH VALUES FOR CO2 EQUILIBRIUM BETWEEN
 3:C      AIR AND WATER
 4:C
 5:       REAL KW,K1,K2
 6:       DIMENSION XPH(100)
 7:       PI=3.14159
 8:       M=5
 9:       N=8
10:C
11:C      PARTIAL PRESSURE OF CO2 IN AIR ASSUMED AS 0.0003 ATM.
12:C
13:       PCO2=0.0003
14:C      INPUT DATA
15:C
16:C      FIRST DATA CARD :NO. OF PH VALUES (NPH) FOR WHICH
17:C      ALKALINITY IS TO BE CALCULATED
18:C
19:       READ(N,10)NPH
20: 10    FORMAT(I2)
21:C
22:C      SECOND SET OF DATA:READ IN PH VALUES
23:C
24:       DO 830 J=1,NPH
25:830    READ(N,840)XPH(J)
26:840    FORMAT(F10.3)
27:       DO 900 II=1,30
28:C
29:C      READ IN IONIC STRENGTH AND TEMPERATURE
30:C
31:       READ(N,800)TS,U
32:800    FORMAT(2F10.4)
33:C
34:C      CALCULATE MONO AND DI VALENT ACTIVETY FACTORS
35:C
36:       U1=U**0.5
37:       FD=2.*((U1/(1+U1))-0.2*U)
38:       FD=10.**FD
39:       FD=1./FD
40:       FM=0.5*((U1/(1+U1))-0.2*U)
41:       FM=10.**FM
42:       FM=1./FM
43:C      CALCULATE EQ. CONSTANTS AT TEMP. TS
```

```
44:         TA=TS+273.
45:         PK1=(17052./TA)+215.21*ALOG(TA)/ALOG(10.)-(.12675*TA)
46:        1-545.56
47:         K1=1./10.**PK1
48:         PK2=(2902.39/TA)+(.02379*TA)-6.498
49:         K2=1./10.**PK2
50:         PKW=(4787.3/TA)+(7.1321*ALOG(TA)/ALOG(10.))+(.01037*TA)

51:        1-22.801
52:         KW=1./10.**PKW
53:C        CONVERT K VALUES TO INCLUDE ACTIVETY COEFFTS.
54:C
55:         K1=K1/FM**2.
56:         K2=K2/FD
57:         KW=KW/FM**2.
58:C
59:C        CONVERT  K VALUES TO PPM AS CACO3
60:C
61:         K1=K1*2.5*10.**4.
62:         K2=K2*10.**5.
63:         KW=KW*2.5*10.**9.
64:         WRITE(M,804)
65:804      FORMAT('ALKALINITY',4X,'PH')
66:C        CALCULATE ALKALINITY FROM  INPUT PH AND CO2 CONC.
67:C        DEFINED BY HENRY'S LAW
68:C
69:C        CALCULATE HENRY'S LAW CONSTANT FOR CO2 AT TEMP. TS
70:C
71:         IF(TS-33.)815,815,820
72:815      PCH=1.12+0.0138*TS
73:         GO TO 825
74:820      PCH=1.36+.0069*TS
75:825      CH=1./10.**PCH
76:C
77:C        CH IS HENRY'S CONSTANT AND CONC.=CH*PARTIAL PR. M
78:C
79:C        EXPRESS  CO2 CONC. IN PPM AS CACO3
80:C
81:         CO2=CH*PCO2
82:         CO2=CO2*10.**5.
83:         WRITE(M,1000)TS,U
84:1000     FORMAT(//,2E11.4)
85:         DO 860 I=1,NPH
86:         PH=XPH(I)
87:         H=1./10.**PH/FM
88:         H=H*5*10.**4.
89:         A=CO2*((K2*K1/H**2.)+(K1/H))+KW/H-H
90:         WRITE(M,810)A,PH
91:810       FORMAT(2F10.4)
92:860      CONTINUE
93:900      CONTINUE
94:         STOP
95:         END
```

NAME INDEX

Arbatsky, J.W. 255

Bard, A.J. 29
Bacon, J.R. 318
Bjerrum, N. 21
Buswell, A.M. 125, 137
Butler, J.N. 57

Caldwell, D.H. 255
Carlson, E.T. 126, 137

Frear, G.L. 125, 137

Ham, R.K. 78
Hamer, P. 84, 125, 126, 245,
    255, 305, 318
Harned, H.S. 84, 137

Jackson, J. 84, 137, 255, 318
Johnston, J. 125, 137

Kleijn, H.F.W. 78

Langlier, W.F. 128, 130-137,
    304, 305, 315
Larson, T.E. 125, 137
Lawrence, W.B. 255
Lawson, S.P. 256, 270, 304
Lill, J.R. 318

MacInnes, D. 84

Nordell, E. 125, 137
Nouwell, T. 126, 137

Peppler, R.B. 126, 137
Plass, G.N. 1
Powell, S.T. 305, 318

Ryznar, J.W. 126, 137

Scholes, S.R. 84, 121
Shadlovsky, T. 84, 121
Sillen, L.G. 21

Snyders, R. 256
Stumm, W. 78, 115, 130, 132, 245

Thurston, E.F. 84, 125, 126, 255,
    318
Travers, A. 126, 137

van Slijke, D.D. 57, 78

Weber, W.J. 78, 115, 245, 255
Wells, L.S. 126, 137

SUBJECT INDEX